PENGUIN BO

Numbers Don't Lie

'There is perhaps no other academic who paints pictures with numbers like Smil' *Guardian*

'An author who does not allow facts to be obscured or overshadowed by politics' *New York Review of Books*

'For anyone confused by statistics or dubious of data in a world where numbers seem to mean everything and nothing' *BBC Science Focus*

'A radical thinker' *Financial Times*

'A slayer of bullshit' David Keith, Harvard University

'In a world of specialized intellectuals, Smil is an ambitious and astonishing polymath who swings for fences' *Wired*

'He is a distinguished professor on the environment faculty at the University of Manitoba but really should be in the department of everything' *The New York Times*

ABOUT THE AUTHOR

Vaclav Smil is Distinguished Professor Emeritus at the University of Manitoba. He is the author of over forty books on topics including energy, environmental and population change, food production and nutrition, technical innovation, risk assessment and public policy. He is a Fellow of the Royal Society of Canada and a Member of the Order of Canada.

Numbers Don't Lie

*71 Things You Need to Know
About the World*

VACLAV SMIL

PENGUIN BOOKS

PENGUIN BOOKS

UK | USA | Canada | Ireland | Australia
India | New Zealand | South Africa

Penguin Books is part of the Penguin Random House group of companies
whose addresses can be found at global.penguinrandomhouse.com.

First published by Viking 2020
Published in Penguin Books 2021

002

Copyright © Vaclav Smil, 2020

The moral right of the author has been asserted

Printed and bound in Great Britain by Clays Ltd, Elcograf S.p.A.

The authorized representative in the EEA is Penguin Random House Ireland,
Morrison Chambers, 32 Nassau Street, Dublin D02 YH68

A CIP catalogue record for this book is available from the British Library

ISBN: 978-0-241-98969-2

www.greenpenguin.co.uk

Contents

CONTENTS

COUNTRIES

Nations in the Age of Globalization

MACHINES, DESIGNS, DEVICES

Inventions That Made Our Modern World

CONTENTS

FUELS AND ELECTRICITY
Energizing Our Societies

TRANSPORT
How We Get Around

vii

FOOD
Energizing Ourselves

CONTENTS

ENVIRONMENT
Damaging and Protecting Our World

Introduction

Numbers Don't Lie is an eclectic book, with topics ranging from people, populations, and countries, through to energy use, technical innovation, and the machines and devices that define our modern civilization. For good measure, it closes with some factual perspectives on our food supply and eating choices, and on the state and degradation of our environment. These are the big topics I have pursued in my books since the 1970s.

First and foremost, this book is about getting the facts straight. But that is not as easy as it might seem: while the World Wide Web teems with numbers, too many of them are undated hand-me-downs of unknown provenence, often with questionable unit identifiers. For example, French GDP in 2010 was US \$2.6 trillion: was that value in current or constant monies, and was the conversion from euros to dollars done using the prevailing exchange rate or purchasing power parity? And how would you know?

In contrast, nearly all the numbers in this book are taken from four kinds of primary sources: worldwide statistics published by global organizations,* yearbooks issued by national institutions,† historical statistics compiled by national agencies,‡ and papers in scientific journals.§ A small share of the numbers come from scientific monographs, from recent studies done by major consulting agencies (known for the reliability of their reports), or from public opinion polls conducted by such long-established organizations as Gallup or the Pew Research Center.

To understand what is really going on in our world, next we must set the numbers in the appropriate contexts: historical and international. For example, starting with the *historical* context, the scientific unit of energy is one joule, and affluent economies now consume annually about 150 billion joules (150 gigajoules) of primary energy per capita (for comparison, one ton of crude oil is 42 gigajoules); while Nigeria, Africa's most populous (and oil- and natural gas–rich) nation, averages only 35 gigajoules. The difference is impressive, with France or Japan using nearly five times as much energy per

* Ranging from the Eurostat and the International Atomic Energy Agency to the UN's World Population Prospects and the World Health Organization.

† My favorite, for their unmatched detail and data quality, are the *Japan Statistical Yearbook* and the USDA's National Agricultural Statistics Service.

‡ Including the exemplary *Historical Statistics of the United States, Colonial Times to 1970* and *Historical Statistics of Japan.*

§ Ranging from *Biogerontology* to the *International Journal of Life Cycle Assessment.*

capita, but the historical comparison illuminates the *real* size of the gap: Japan used that much energy by 1958 (an African lifetime ago), and France averaged 35 gigajoules already by 1880, putting Nigeria's access to energy *two* lifetimes behind France.

Contemporary *international* contrasts are no less memorable. Comparing the American infant mortality rate with that of sub-Saharan Africa reveals a large but expected gap. And that the United States does not belong to the top 10 countries with the lowest infant mortality is not that surprising considering its highly diverse population and high rates of immigration from less developed countries—but few people would guess that it does not rank even among the top 30 nations!* This surprise leads, inevitably, to asking why this is the case, and that question opens a universe of social and economic considerations. True comprehension of many numbers (individually or part of complex statistics) requires a combination of basic scientific literacy and numeracy.

Length (distance) is the easiest measure to internalize. Most people have a fair grasp of 10 centimeters (the width of an adult fist with a thumb held outside), a meter (roughly ground-to-waist for an average man), and a kilometer (a one-minute drive in city traffic). Common speeds (distance/time) are also easy: a brisk walk is 6 km/h, a rapid intercity train is 300 km/h, a jetliner

* In 2018, it was 33rd out of 36 OECD nations.

pushed by a strong jet stream does 1,000 km/h. Masses are more difficult to "feel": a newborn child usually weighs less than 5 kilograms, a small deer less than 50 kilograms, some battle tanks weigh less than 50 tons, and the maximum takeoff weight of an Airbus 380 is more than 500 tons. Volumes can be equally tricky: the gasoline tank of a small sedan is less than 40 liters; the internal volume of a small American house is usually less than 400 cubic meters. Getting the feel for energy and power (joules and watts) or current and resistance (amperes and ohms) is hard without working with these units frequently—so relative comparisons, such as the gap between African and European energy use, are easier.

Money presents different challenges. Most people appreciate relative levels of their incomes or savings, but *historic* comparisons on a national and international level must be adjusted for inflation, and *international* comparisons must account for fluctuating exchange rates and changing purchasing powers.

And then there are qualitative differences that cannot be captured by numbers, and such considerations are particularly important when comparing food preferences and diets. For example, carbohydrate and protein content per 100 grams may be very similar, but what passes for bread in an Atlanta supermarket (pre-sliced soft square shapes packed in plastic sleeves) is—quite literally—an ocean apart from what a *maître boulanger* or a *Bäckermeister* would display in their shops in Lyon or Stuttgart.

As numbers get bigger, orders of magnitude (tenfold differences) become more telling than the specific figures: an Airbus 380 is an order of magnitude heavier than a battle tank; a jetliner is an order of magnitude faster than a car on a freeway; and a deer weighs an order of magnitude more than a baby. Or, using superscripts and multiples according to the International System of Units, a newborn baby is 5×10^3 grams or 5 kilograms; an Airbus 380 is more than 5×10^8 grams or 500 million grams. As we get into *really* big numbers, it does not help that Europeans (taking the French lead) deviate from scientific notation and do not call 10^9 a billion but (*vive la différence!*) *un milliard* (resulting in *une confusion fréquente*). The world will soon have 8 billion people (8×10^9), in 2019 its economic output (in nominal terms) was about \$90 trillion (\$$9 \times 10^{13}$), and it consumed more than 500 billion billions of joules of energy (500×10^{18}, or 5×10^{20}).

The good news is that mastering much of this is easier than most people think. Suppose you put down your mobile phone (I have never owned one, nor felt like I was missing out) for a few minutes a day, and estimated the lengths and distances around you—checking them, perhaps with your fist (remember, about 10 centimeters) or (after picking your mobile back up) through GPS. You should also try to calculate the volumes of objects you encounter (people always underestimate the volume of thin but large objects), and it is an outright entertaining proposition to calculate (without any electronic help) the differences in orders of magnitude as you read about the latest national income inequalities between billion-aires and Amazon warehouse packers (how many orders of magnitude separate their annual take?), or as you see a comparison of average national per capita GDP rates (how many orders of magnitude is the United Kingdom above Uganda?). These mental exercises will put you in touch with the physical realities of the surrounding world while keeping your synapses firing. Understanding numbers simply takes a bit of engagement.

My hope is that this book will help readers understand the true state of our world. I hope it might surprise you, cause you to marvel at the uniqueness of our species, at our inventiveness and our pursuit of better understanding. My goal is to demonstrate not only that numbers do not lie, but to discover which truth they convey.

One final note on the numbers contained within—all dollar amounts, unless otherwise specified, are in US $; and all measurements are given in metric form, with a few warranted exceptions such as nautical miles and inches for American lumber.

Vaclav Smil
Winnipeg, 2020

PEOPLE
The Inhabitants of Our World

What happens when we have fewer children?

Total fertility rate (TFR) is the number of children born per woman during her lifetime. The most obvious physical constraint on this is the length of the fertile period (from menarche to menopause). The age of first menstruation has been decreasing from about 17 years in preindustrial societies to less than 13 years in today's Western world, while the average onset of menopause has advanced slightly, to just above 50, resulting in a typical fertile span of some 38 years compared to about 30 years in traditional societies.

There are 300–400 ovulations during the fertile lifespan. As every pregnancy precludes 10 ovulations and because an additional 5–6 ovulations have to be subtracted for each pregnancy, due to the reduced chance of conception during the traditionally prolonged breastfeeding period, the maximum fertility rate is about two dozen pregnancies. With some multiple births the total can be in excess of 24 live births, confirmed by historical records of women having more than 30 children.

But typical maximum fertility rates in societies practicing no birth control have always been much lower than this, due to the combination of pregnancy loss, still births, infertility, and premature maternal mortality.

Rapid fertility rate declines in Asia compared with a steady Africa

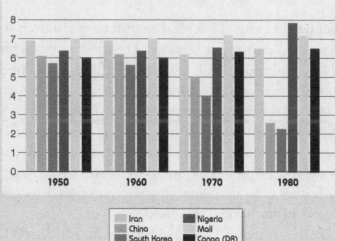

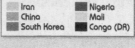

Iran
China
South Korea
Nigeria
Mali
Congo (DR)

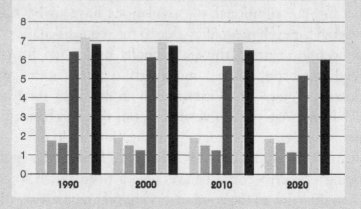

These realities reduce maximum population-wide fertilities to 7–8; indeed, such rates were common on all continents well into the 19th century, in parts of Asia until two generations ago, and they can still be found in sub-Saharan Africa, with Niger at 7.5 (which is well below the preferred family size: when asked, the average number of children that Nigerien women would prefer is 9.1!). But even in that region, the TFR—although still high—has declined (to 5–6 in most of those countries), and the rest of the world now lives with moderate, low, and extremely low fertilities.

The transition toward this new world began at different times, not only among different regions but also within regions: France was far ahead of Italy, Japan far ahead of China—and Communist China eventually took the drastic step of restricting families to a single child. That aside, desire for fewer children has been driven by an often highly synergistic combination of gradually rising standards of living, the mechanization of agricultural work, the displacement of animals and people by machines, mass-scale industrialization and urbanization, increasing numbers of females in the urban labor force, advancing universal education, better healthcare, a higher survival rate of newborns, and government-guaranteed pensions.

A historic pursuit of quantity turned, sometimes rapidly, into the quest for quality: the benefits of high fertility (ensuring survival amid high infant mortality; supplying additional labor force; providing old-age insurance) began to weaken and then to disappear, and smaller families invested more in their children and in raising their quality of life, usually starting with better nutrition (more meat and fresh fruit; more eating out) and ending with SUVs and flights to distant tropical beaches.

As is often the case in both social and technical transitions, the pathbreakers took a long time to accomplish the change, while some late adopters completed the process in just two generations. The shift from high to low fertility took about two centuries in Denmark and about 170 years in Sweden. In contrast, South Korean fertility fell from more than 6 TFR to below the replacement level in just 30 years, and even before the introduction of the one-child policy, Chinese fertility had plunged from 6.4 in 1962 to 2.6 in 1980. But the unlikely record holder is Iran. In 1979, when the monarchy was overthrown and Ayatollah Khomeini returned from exile to establish a theocracy, Iran's fertility averaged 6.5, but by the year 2000 it was down to the replacement level and it has continued to fall.

The replacement level of fertility is that which maintains a population at a stable level. It is about 2.1, with the additional fraction needed to make up for girls who will not survive into fertile age. No country has been able to stop the fertility decline at the replacement level and achieve a stationary population. An increasing share of humanity lives in societies with below-replacement fertility levels. In 1950, 40 percent of humanity lived in countries with fertilities above 6 and the mean rate was about 5; by the year 2000, just 5 percent of the global population was in countries with fertilities above 6, and the mean (2.6) was close to the replacement level. By 2050, nearly three-quarters of humanity will reside in countries with below-replacement fertility.

This nearly global shift has had enormous demographic, economic, and strategic implications. European importance has diminished (in 1900 the continent had about 18 percent of the world's population; in 2020 it has only 9.5 percent) and Asia has ascended (60 percent of the world total in 2020), but regional high fertilities guarantee that nearly 75 percent of all births during the 50 years between 2020 and 2070 will be in Africa.

And what does the future hold for countries whose fertility has fallen below the replacement level? If the national rates remain close to the replacement (no lower than 1.7; France and Sweden were at 1.8 in 2019), then there is a good chance of possible future rebounds. Once they slip below 1.5, such reversals appear increasingly unlikely: in 2019, there were record lows of 1.3 in Spain, Italy, and Romania, and 1.4 in Japan, Ukraine, Greece, and Croatia. Gradual population decline (with all of its social, economic, and strategic implications) seems to be the future of Japan and of many European countries. So far, no pro-natalist government policies have brought any major reversal, and the only obvious option to prevent depopulation is to open the gates for immigration—but that looks unlikely to happen.

The best indicator of quality of life? Try infant mortality

When looking for the most revealing measures of human quality of life, economists—ever ready to reduce everything to money—prefer to rely on per capita values of gross domestic product (GDP) or disposable income. Both measures are obviously questionable. GDP goes up in a society where increasing violence requires more policing, higher investment in security measures, and more frequent admissions to hospitals; and average disposable income tells us nothing about the degree of economic inequality or about the social net available to disadvantaged families. Even so, these measures do give a pretty good overall ranking of countries. Not too many people would prefer to live in Iraq (2018 nominal GDP of about $6,000) than in Denmark (2018 nominal GDP of about $60,000). And average quality of life is undoubtedly higher in Denmark than in Romania: both belong to the EU, but the disposable income is 75 percent higher in the former.

Since 1990, the most common alternative has been to use the Human Development Index (HDI), a multivariable measure constructed in order to provide a better yardstick. It combines life expectancy at birth and educational achievements (mean and expected years of

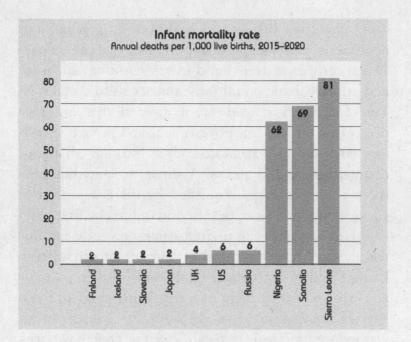

Infant mortality rate
Annual deaths per 1,000 live births, 2015–2020

schooling) with the gross national income per capita—but (not surprisingly) it correlates highly with the average per capita GDP, making the latter variable about as good a measure of the quality of life as the more elaborate index.

My own choice of a single-variable measure for rapid and revealing comparisons of quality of life is infant mortality: the number of deaths during the first year of life that take place per 1,000 live births.

Infant mortality is such a powerful indicator because low rates are impossible to achieve without having a combination of several critical conditions that define good quality of life—good healthcare in general, and appropriate prenatal, perinatal, and neonatal care in particular; proper maternal and infant nutrition; adequate and sanitary living conditions; and access to social support for disadvantaged families—and that are also predicated on relevant government and private spending, and on infrastructures and incomes that can maintain usage and access. A single variable thus captures a number of prerequisites for the near-universal survival of the most critical period of life: the first year.

Infant mortalities in preindustrial societies were uniformly and cruelly high: even by 1850 the rates in western Europe and in the United States were as high as 200–300 (that is, every fifth to every third child did not survive the first 365 days). By 1950, the Western mean was reduced to 35–65 (typically one out of 20 newborns died within its first year), and now the lowest rates in affluent countries are below 5 (with one infant among 200 not seeing its first birthday). After leaving out minuscule jurisdictions—from Andorra and Anguilla to Monaco and San Marino—this group with infant mortalities lower than 5 per 1,000 includes about 35 countries ranging from Japan (at 2) to Serbia (at just under 5), and its frontrunners show why the measure cannot be used for simplistic ranking without reference to wider demographic conditions.

Countries with the lowest infant mortalities are mostly small (with populations less than 10 million and usually less than 5 million), they include the world's most homogeneous societies (Japan and South Korea in Asia; Iceland, Finland, and Norway in Europe), and most of them have very low birth rates. Obviously, it is more challenging to reach and maintain very low infant mortalities in larger, heterogeneous societies with high rates of immigration from less affluent countries, and in countries with higher birth rates. As a result, it would be difficult to replicate the Icelandic rate (3) in Canada (infant mortality at 5), a country whose population is more than 100 times larger and that welcomes annually about as many newcomers (from scores of countries, and mostly from low-income societies in Asia) as there are total people living in Iceland. The same realities affect the United States, but the country's relatively high infant mortality (6) is undoubtedly influenced (as is, to a lesser degree, the Canadian rate) by higher economic inequality.

In this sense, infant mortality is a more discerning indicator of quality of life than the income average or the Human Development Index, but it still needs qualifications: no single measure is a fully satisfactory proxy for gauging a nation's quality of life. What is not in doubt is that infant mortalities remain unacceptably high in a dozen of sub-Saharan nations. Their rates (above 60 per 1,000) are equal to those in western Europe some 100 years ago, a timespan that evokes the developmental gap those nations have to close in order to catch up with affluent economies.

The best return on investment: Vaccination

Death due to infectious diseases in infancy and child-hood remains perhaps the cruelest fate in the modern world, and one of the most preventable. Measures needed to minimize this untimely mortality cannot be ranked as to their importance: clean drinking water and adequate nutrition are as vital as disease prevention and proper sanitation. But if you judge them by their benefit-cost ratios, vaccination is the clear winner.

Modern vaccination dates back to the 18th century, when Edward Jenner introduced it against smallpox. Vaccines against cholera and plague were created before the First World War, and others against tuberculosis,

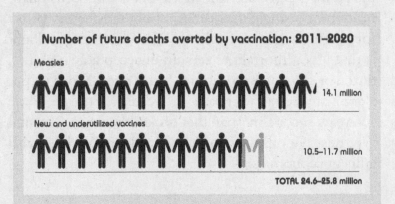

Number of future deaths averted by vaccination: 2011–2020

Measles

14.1 million

New and underutilized vaccines

10.5–11.7 million

TOTAL 24.6–25.8 million

tetanus, and diphtheria before the Second. The great postwar breakthroughs included routine vaccinations against pertussis (whooping cough) and polio. Today, the standard practice everywhere is to inoculate children with a pentavalent vaccine that prevents diphtheria, tetanus, pertussis, and polio, as well as meningitis, otitis, and pneumonia, three infections caused by *Haemophilus influenzae* type B. The first dose comes 6 weeks after birth; the other two follow at 10 and 14 weeks. Each pentavalent vaccine costs less than $1, and every additional vaccinated child reduces the chances of infection among unvaccinated peers.

Given these realities, it has been always clear that vaccination has an extraordinarily high benefit-cost ratio, though one that is not easy to quantify. But thanks to a 2016 study supported by the Bill & Melinda Gates Foundation and conducted by US healthcare professionals in Baltimore, Boston, and Seattle, we can finally measure the payoff. The study's focus was on the return on investment associated with projected vaccination coverage levels in nearly 100 low- and middle-income countries during the second decade of this century— the decade of vaccines.

Benefit-cost ranges were based on the costs of vaccines and of their supply and delivery chains on one hand, and on estimates of the averted costs of morbidity and mortality on the other. For every dollar invested in vaccination, $16 is expected to be saved in healthcare costs and the lost wages and lost productivity caused by illness and death.

And when the analysis went beyond the restricted cost-of-illness approach and looked at broader economic benefits, it found the net benefit-cost ratio was more than twice as high—reaching 44 times, with an uncertainty range of 27 to 67. The highest rewards were for averting measles: a 58-fold return.

The Gates Foundation released the finding of the 44-fold benefit in the form of a letter to Warren Buffett, the foundation's largest outside donor. Even he must be impressed with such a return on investment!

There is still some way to go. After generations of progress, the basic vaccination coverage in high-income countries is now nearly universal, at around 96 percent, and great advances have been made in low-income nations, where the coverage has risen from only 50 percent in the year 2000 to 80 percent in 2016.

The hardest part might be to eliminate the threat of infectious diseases entirely. Polio is perhaps the best illustration of this challenge: the worldwide infection rate dropped from some 400,000 cases in 1985 to fewer than 100 by the year 2000, but in 2016 there were still 37 cases in violence-beset regions of northern Nigeria, Afghanistan, and Pakistan. And, as illustrated recently by the Ebola, Zika, and COVID-19 viruses, new infection risks will arise. Vaccines remain the best way to control them.

Why it's difficult to predict how bad a pandemic will be while it is happening

I wrote the first version of this chapter at the end of March 2020, just as the COVID-19 pandemic was going through its first exponential rise throughout Europe and in North America. Rather than offering yet another estimate or prediction (and hence making the chapter instantly obsolete), I decided I would explain the uncertainties that always complicate our judgment and our interpretation of statistics in these stressful situations.

Fears engendered by a viral pandemic are due to relatively high mortalities, but it is impossible to pinpoint those rates while the infection is spreading—and it is difficult to do so even after a pandemic ends. The most common epidemiological approach is to calculate the case fatality risk: confirmed deaths associated with a virus are divided by the number of cases. The numerator (death certificates stating the cause of mortality) is obvious and in most countries that count is fairly reliable. But the choice of the denominator introduces many uncertainties. Which "cases"? Only laboratory-confirmed infections, all symptomatic cases (including people who were not tested but displayed expected symptoms), or

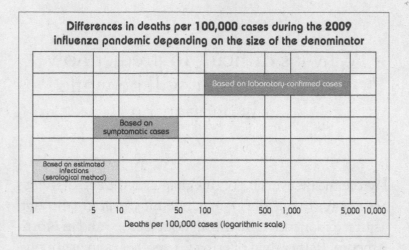

Differences in deaths per 100,000 cases during the 2009 influenza pandemic depending on the size of the denominator

Based on laboratory-confirmed cases

Based on symptomatic cases

Based on estimated infections (serological method)

1   5   10   50   100   500   1,000   5,000   10,000

Deaths per 100,000 cases (logarithmic scale)

the total number of infections including asymptomatic cases? Tested cases are known with high accuracy, but the total number of infections must be estimated either by relying on post-pandemic population serological studies (finding antibodies in blood), by using various growth equations to model the past spread of a pandemic, or by assuming the most likely multipliers (x people infected for y people who actually died).

Detailed study of case fatalities of the 2009 influenza pandemic—which began in the US in January 2009, lingered in some places until August 2010, and was caused by a new H1N1 virus—illustrates the range of resulting uncertainties. Confirmed deaths were always in the numerator, but for the denominator there were three different categories of case definition: laboratory-confirmed cases, estimated symptomatic cases, and estimated infections (based on serology or on assumptions regarding the extent of asymptomatic infections). Resulting differences were very large, from less than 1 to more than 10,000 deaths per 100,000 people.

As expected, the laboratory-confirmed approach yielded the highest risk (mostly between 100 and 5,000 deaths), while the symptomatic approach had the range of 5–50 deaths, and the estimated infections in the denominator yielded risks of just 1–10 deaths per 100,000 people: the first approach showed fatalities being up to 500 times higher than the last one did!

In 2020, with the spread of COVID-19 (caused by a coronavirus, SARS-CoV-2), we face the same uncertainties. The COVID-19 pandemic began in Wuhan, capital of China's Hubei province, in late 2019. By March 30, 2020, when the worst seemed to be over, official Chinese statistics listed 50,006 cases in the city and 2,547 deaths. On April 17 the Chinese raised the death toll by a bit more than 50 percent to 3,869—but no new deaths were recorded by November 2020 while the cases rose only marginally to 50,340. There was no independent confirmation of these suspect totals, and it is unlikely that we will ever know the real numbers. Official numbers imply that less than 0.5 percent of 11.1 million Wuhan inhabitants were infected, an incredibly low share compared to the numbers of people affected by annual flu, but that the case mortality was fairly high at 7.7 percent.

Interim US numbers show that by November 11, eight months after the WHO-designated start of the pandemic,

COVID-19 case mortality and overall mortality were already far higher than those caused by seasonal flu—and they were still rising at record rates. The Centers for Disease Control and Prevention (CDC) estimated that in the US the 2019–2020 relatively mild seasonal flu infected 38 million Americans (out of a population of about 330 million), and that it resulted in 22,000 deaths. This means that nearly 12 percent of all Americans got infected, and that nearly 0.06 percent of all infected people died (case mortality rate); the overall influenza-specific mortality would be 0.07/1,000 (that is, less than one person among every 10,000 people dying). By November 11, 2020, about 10.5 million Americans (just over 3 percent of the population) had been infected by SARS-CoV-2, and 245,000 died. This means that the COVID-19 case mortality rate (2.34 percent) would be about forty times that of the 2019–2020 seasonal flu case mortality, while the overall COVID-specific mortality would be 0.74 percent, or about eleven times the seasonal flu rate. The overall number of COVID-19-related deaths (excess mortality, deaths above the normally expected total) was even higher—but, as with every pandemic, we will have to wait until COVID-19 runs its course to get a clear picture of how bad it was.

Only then will we be able to do the actual counts—or, because we may never know the total number of infected people nationally and globally, simply offer our best estimates—and compare the resulting case fatality risks, which may differ no less than the numbers for the

2009 pandemic. This is one of the most basic algebraic lessons: you may know the exact numerator, but unless you know the denominator with a comparable certainty, you cannot calculate the precise rate. Uncertainties will never fully go away, but by the time you read this, we will have a much better understanding of the true extent and intensity of the latest pandemic than when these lines were written during the first (March) and the second (November) wave of the global spread. I trust you'll still be reading.

Growing taller

Like many other inquiries into the human condition, studies of human height have their belated origins in 18th-century France, where between 1759 and 1777 Philibert Guéneau de Montbeillard measured his son every six months—from birth to his 18th birthday—and the Comte de Buffon published the table of the boy's measurements in the 1777 supplement to his famous *Histoire Naturelle*. But Montbeillard's son was tall for his time (as a young adult he matched today's average Dutchman), and we didn't see systematic large-scale data on human height and the growth of children and adolescents until the 1830s, courtesy of Edouard Mallet and Adolphe Quetelet.

Since then we have studied all aspects of human height, ranging from its expected progress with age and its relation to weight, to its nutritional and genetic determinants and gender differences in growth spurts. As a result, we know—with high accuracy—expected heights (and weights) at different ages. If a young American mother comes to her pediatrician with a two-year-old boy measuring 93 centimeters, she will be told that her son is taller than 90 percent of children in his cohort.

For those interested in long-term measures of progress,

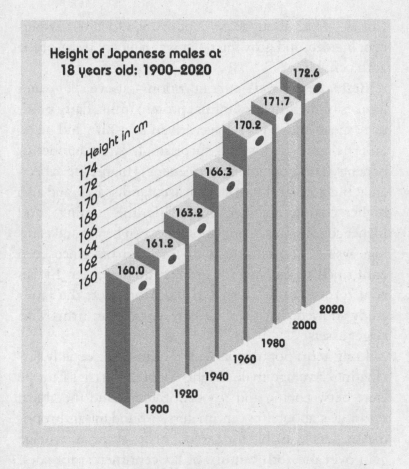

Height of Japanese males at 18 years old: 1900–2020

Height in cm

174
172
170
168
166
164
162
160

160.0 — 1900
161.2 — 1920
163.2 — 1940
166.3 — 1960
170.2 — 1980
171.7 — 2000
172.6 — 2020

as well as revealing international comparisons, one of the best outcomes of modern systematic studies of growth has been the well-documented history of rising average heights. Although stunting (inadequate growth that produces low height for age in young children) remains common in many poor countries, its global prevalence has declined—mostly thanks to rapid improvement in

China—from about 40 percent in 1990 to about 22 percent in 2020, and growing taller was a global trend of the 20th century.

Better health and better nutrition—above all, greater intakes of high-quality animal protein (milk, dairy products, meat, and eggs)—have driven the shift, and being taller is associated with a surprisingly large number of benefits. These do not include generally higher life expectancy, but a lower risk of cardiovascular diseases, and also higher cognitive ability, higher lifetime earnings, and higher social status. Correlation between height and earnings was first documented in 1915 and has since been confirmed repeatedly, for groups ranging from Indian coal miners to Swedish CEOs. Moreover, the latter study showed that the CEOs were taller in firms with larger assets!

Long-term population-wide findings are equally fascinating. Average male heights in preindustrial Europe were between 169 and 171 centimeters, and the global mean was about 167 centimeters. Abundant anthropometric data available for 200 countries show an average gain over the 20th century of 8.3 centimeters for adult women and 8.8 for men. The population of every country in Europe and North America got taller, while South Korean women recorded the century's largest average female gains (20.2 centimeters) and Iranian men topped the male sequence with 16.5 centimeters. Detailed Japanese data, recorded since 1900 for both sexes at 12 different ages between 5 and 24, show how growth responds to nutritional constraints and improvements: between 1900 and 1940 the average height of 10-year-old boys rose by 0.15 cm/year, but wartime food shortages cut it by 0.6 cm/year; the annual increase resumed only in 1949, and during the second half of the century it averaged 0.25 cm/year. Similarly, the Chinese gains were interrupted by the world's largest famine (1959–61), but males in major cities still averaged an increase of 0.13 cm/year during the latter half of the 20th century. In contrast, measurements for the second half of the 20th century show minimal gains in India and Nigeria, none in Ethiopia, and a slight decline in Bangladesh.

And which nation has the tallest citizens? For males the record holders are the Netherlands, Belgium, Estonia, Latvia, and Denmark; for females it is Latvia, the Netherlands, Estonia, the Czech Republic, and Serbia; and the tallest cohort (whose average surpasses 182.5 centimeters) is that of the Dutchmen born during the last quarter of the 20th century. Milk has been a key growth factor, be it in Japan or in the Netherlands. Before the Second World War, Dutch males were smaller than American men, but post-1950 US milk consumption declined while in the Netherlands it rose until the 1960s—and it remains higher than in the US. The lesson is obvious: the easiest way to improve a child's chances of growing taller is for them to drink more milk.

Is life expectancy finally topping out?

Ray Kurzweil, Google's chief futurist, says that if you can just hang on until 2029, medical advances will start to "add one additional year, every year, to your life expectancy. By that I don't mean life expectancy based on your birthdate, but rather your remaining life expectancy." Curious readers can calculate what this trend would do to the growth of the global population, but I will limit myself here to a brief review of survival realities.

In 1850, the combined life expectancies of men and

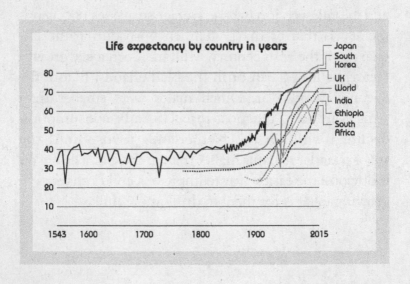

women stood at around 40 years in the United States, Canada, Japan, and much of Europe. Since then, the values have followed an impressive and almost perfectly linear increase that saw them nearly double. Women live longer in all societies, with the current maximum at just above 87 years in Japan.

The trend may well continue for a few decades, given that from 1950 to 2000 the life expectancies of elderly people in affluent countries rose at about 34 days per year. But without fundamental discoveries that change the way we age, this trend to longer life must weaken and finally end. The long-term trajectory of Japanese female life expectancy—which increased from 81.91 years in 1990 to 87.26 years in 2017—fits a symmetrical logistic curve that is already close to its asymptote of about 90 years. The trajectories for other affluent countries also show the approaching ceiling. Records available for the 20th century show two distinct periods of rising longevity: faster linear gains (about 20 years in half a century) prevailed until 1950, followed by slower gains.

If we are still far from the limit to the human lifespan, then the largest survival gains should be recorded among the oldest people, i.e., 80–85 year olds should be gaining more than those who are 70–75 years old. This was indeed the case for studies conducted in France, Japan, the United States, and the United Kingdom from the 1970s to the early 1990s. Since then, however, the gains have leveled off.

There may be no specific genetically programmed

limit to lifespan—much as there is no genetic program that limits us to a specific running speed (see HOW SWEATING IMPROVED HUNTING, p. 28). But lifespan is a bodily characteristic that arises from the interaction of genes with the environment. Genes may themselves introduce biophysical limits, and so can environmental effects such as smoking.

The world record lifespan is the 122 years claimed for Jeanne Calment, a Frenchwoman who died in 1997. Strangely, after more than two decades, she still remains the oldest survivor ever, and by a substantial margin. (Indeed, the margin is so big as to be suspicious; her age and even her identity are in question.) The second-oldest supercentenarian died at 119, in 1999, and since that time there have been no survivors beyond the 117th year.

And if you think that you have a high chance to make it to 100 because some of your ancestors lived that long, you should know that the estimated heritability of life-span is modest, between 15 and 30 percent. Given that people tend to marry others like themselves—a phenomenon known as assortative mating—the true heritability of human longevity is probably even lower than that.

Of course, as with all complex matters, there is always room for different interpretations of published statistical analyses. Kurzweil hopes that dietary interventions and other tricks will extend his own life until major scientific advances can preserve him forever. It is true that there are ideas on how such preservation might be achieved, among them the rejuvenation of human cells by extending their telomeres (the nucleotide sequences at the ends of a chromosome that fray with age). If it works, maybe it can lift the realistic maximum to well above 125 years.

But for now, the best advice I can give to all but a few remarkably precocious readers is to plan ahead—though perhaps not as far ahead as the 22nd century.

How sweating improved hunting

Before the development of long-range projectile weaponry some tens of thousands of years ago in Africa, our ancestors had only two ways to secure meat: by scavenging the leftovers of mightier beasts or by running down their own prey. Humans were able to occupy the second of those ecological niches thanks, in part, to two great advantages of bipedalism.

The first advantage is in how we breathe. A quadruped can take only a single breath per locomotive cycle,

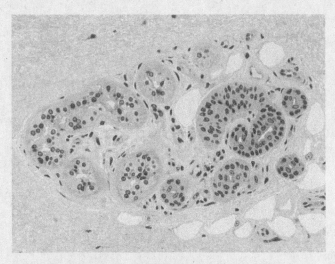

Microscopic section of human eccrine glands

because its chest must absorb the impact on the front limbs. We, however, can choose other ratios, and that lets us use energy more flexibly. The second (and greater) advantage is in our extraordinary ability to regulate our body temperature, which allows us to do what lions cannot: run long and hard in the noonday sun.

It all comes down to sweating. The two large animals we have mainly used for transport perspire profusely compared to other quadrupeds: in one hour, a horse can lose about 100 grams of water per square meter of skin, and a camel can lose up to 250 g/m^2. However, a human being can easily shed 500 g/m^2, enough to remove between 550 and 600 watts' worth of heat. Peak hourly sweating rates can surpass 2 kilograms per square meter, and the highest reported short-term sweating rate is twice that high.

We are the superstars of sweating, and we need to be. An amateur running the marathon at a slow pace will consume energy at a rate of 700–800 watts, and an experienced marathoner who covers the 42.2 kilometers in 2.5 hours will metabolize at a rate of about 1,300 watts.

And we have another advantage when we lose water: we don't have to make up the deficit instantly. Humans can tolerate considerable temporary dehydration providing that we rehydrate in a day or so. In fact, the best marathon runners drink only about 200 milliliters per hour during a race.

Together, these advantages allowed our ancestors to become unrivaled as a diurnal, high-temperature predator. They could not outsprint an antelope, of course, but during a hot day they could dog its heels until it finally collapsed, exhausted.

Documented cases of such long-distance chases come from three continents and include some of the fleetest quadrupeds. In North America, the Tarahumara of north-western Mexico could outrun deer. Further north, Paiutes and Navajos could exhaust pronghorns. In South Africa, Kalahari Basarwa ran down a variety of antelopes and even wildebeests and zebras during the dry season. In Australia, some Aborigines would outrun kangaroos.

These runners would even have had an advantage over the modern runners using expensive athletic shoes: their barefoot running not only reduced their energy costs by about 4 percent (a nontrivial advantage on long runs), it also exposed them to fewer acute ankle and lower-leg injuries.

In the race of life, we humans are neither the fastest nor the most efficient. But thanks to our sweating capability, we are certainly the most persistent.

How many people did it take to build the Great Pyramid?

Given the time elapsed since the completion of Khufu's Great Pyramid (nearly 4,600 years), the structure—albeit stripped of the smooth white limestone cladding that made it shine from afar—stands remarkably intact, and hence there is no argument about its exact shape (a polyhedron with regular polygon base), its original height (146.6 meters including the lost pyramidion or capstone), and volume (about 2.6 million cubic meters).

The great pyramids of Giza

However, we may never know how it was built, because every common explanation is problematic. A single long ramp would have required an enormous amount of material to construct, and moving stones up shorter, wraparound ramps would have been tricky—as would lifting and jacking up more than 2 million stones into position. But just because we do not know how it was erected does not mean that we cannot say with some confidence how many people were required to build it.

We must start with the time constraint of two decades, the length of Khufu's reign (he died around 2530 BCE). Herodotus, writing more than 21 centuries after the pyramid's completion, was told during his visit to Egypt that labor gangs totaling 100,000 men at a time worked in three-month spells to finish the structure. In 1974, Kurt Mendelssohn, a German-born British physicist, put the labor force at 70,000 seasonal workers and up to 10,000 permanent masons. But these are large overestimates, and we can come close to the real number by resorting to inescapable physics.

The Great Pyramid's potential energy (what is required to lift the mass above ground level) is about 2.4 trillion joules. Calculating this is fairly easy: it is simply the product of the acceleration due to gravity, the pyramid's mass, and its center of mass (a quarter of its height). Though the mass cannot be pinpointed because it depends on the specific densities of the Tura limestone and mortar used to build the structure, I am assuming a mean of 2.6 tons per cubic meter and hence a total mass of about 6.75 million tons.

People are able to convert about 20 percent of food energy into useful work, and for hard-working men that amounts to about 440 kilojoules a day. Lifting the stones would thus require about 5.5 million labor days (2.4 trillion divided by 440,000), or about 275,000 days a year during the 20-year period, and about 900 people could deliver that by working 10 hours a day for 300 days a year. A similar number of workers might be needed to emplace the stones in the rising structure and then smooth the cladding blocks (in contrast, many interior blocks were just rough-cut). And in order to cut 2.6 million cubic meters of stone in 20 years, the project would have required about 1,500 quarrymen working 300 days per year and producing 0.25 cubic meters of stone per capita by using copper chisels and dolerite mallets. The grand total of the construction labor would then be some 3,300 workers. Even if we were to double that in order to account for designers, organizers, and overseers, and for the labor needed for transport, repair of tools, the building and maintaining of on-site housing, and cooking and clothes-washing, the total would be still fewer than 7,000 workers.

During the time of the pyramid's construction, the total population of Egypt was 1.5–1.6 million people, and hence the deployed force of less than 10,000 would not have amounted to any extraordinary imposition on the country's economy. The challenge would have been to organize the labor; to plan an uninterrupted supply of building stones, including the granite for internal structures (particularly the central chamber and the massive corbeled grand gallery) that had to be delivered by boats from southern Egypt some 800 kilometers from Giza; and to provide housing, clothing, and food for labor gangs on-site.

In the 1990s, archeologists uncovered a cemetery for workers as well as the foundations of a settlement used to house the builders of the two later pyramids at Giza, indicating that no more than 20,000 people lived at the site. The rapid sequence of building two additional pyramids (for Khafre, Khufu's son, starting in 2520 BCE; and for Menkaure, starting in 2490 BCE) is the best testimony to the fact that pyramid-building had been mastered to such a degree that the erection of those massive structures became just another set of construction projects for the Old Kingdom's designers, managers, and workers.

Why unemployment figures do not tell the whole story

Many economic statistics are notoriously unreliable, and the reason often has to do with what is included in the measurement and what is left out. Gross domestic product offers a good example of a measure that leaves out key environmental externalities such as air and water pollution, soil erosion, biodiversity loss, and the effects of climate change.

Unemployed men lining up for food during the Great Depression

Measuring unemployment is also an exercise in exclusion, and the choices are perhaps best illustrated with detailed data from the United States. Casual consumers of US economic news will be familiar only with the official figure, which put the country's total unemployment at 3.5 percent in December 2019. But that is just one of six different methods used by the Bureau of Labor Statistics to quantify "labor underutilization."

Here they are, in ascending order (with rates, again, for December 2019). People unemployed 15 weeks or longer as a share of the civilian labor force: 1.2 percent. People who lost jobs and who completed temporary jobs: 1.6 percent. Total unemployment as a share of the civilian labor force (the official rate): 3.5 percent. Total unemployed plus discouraged workers (those no longer looking for a job), as a share of the civilian labor force and discouraged workers: 3.7 percent. The previous category enlarged by all people only "marginally attached" (doing temporary or occasional work) to the labor force: 4.2 percent. And finally, the last category plus those who work only part-time for economic reasons (that is, they would prefer to work full-time): 6.7 percent. These six measures present quite a spread of values—the official unemployment rate (U-3) was only about half of the most encompassing rate (U-6), which was more than five times as high as the narrowest measure (U-1).

If you lose your job, you count as unemployed only if you keep looking for a new one; otherwise, you never get counted again. That is why, when trying to get closer to the "real" unemployment rate, you must look at the labor force participation rate (the number of people available for work as a percentage of the total population), which has recently been in decline. In 1950 the US rate was only about 59 percent, and after mostly rising for half a century it peaked at 67.3 percent during the spring of 2000; the subsequent decline brought it to 62.5 percent by the fall of 2005, and this was followed by a slow rise to 63.2 by the end of 2019. There are, of course, substantial differences among age groups: the highest rate is about 90 percent, for men between 35 and 44 years of age.

And European unemployment figures show how difficult it is to relate them to a country's social fabric or to its inhabitants' personal satisfaction. The lowest rate, at just above 2 percent, is in the Czech Republic; while Spain has endured years of high unemployment—more than 26 percent in 2013 and more than 14 percent in late 2019 for the entire population, and, even after declining a bit, still about 33 percent for Spanish youth in 2019 (the latter figure clearly a depressing reality for anybody entering the labor force). And yet the Czech happiness score (see the following chapter) is only 8 percent ahead of the Spanish one, and the Czech suicide rate is, at just over 8 per 100,000, three times as high as in Spain. True, robberies are more common in Barcelona than in Prague, but the Spanish mean is only slightly higher than the British mean—and British unemployment is a quarter of the Spanish rate.

Obviously, complex realities of (un)employment can never be caught by an aggregate number. Many people who have been formally unemployed have coped thanks to family support and informal labor arrangements. Many who are fully employed are unhappy with their lot but cannot change jobs easily or at all, because of their skills or family circumstances. Numbers may not lie, but individual perceptions of them differ.

What makes people happy?

To answer that question, it would be very helpful to know which societies actually consider themselves significantly happier than others—and, since 2012, this is as easy as consulting the latest edition of the *World Happiness Report*, now published annually in New York by the United Nations Sustainable Development Solutions Network. In 2019 (summing up data and surveys from 2016–18), Finland was the world's happiest country for the second time in a row, followed by Denmark, Norway, and Iceland; the Netherlands and Switzerland came just ahead of Sweden, which means that the Nordic nations took five of the top seven spots. The top 10 were rounded out by New Zealand, Canada, and Austria. The second group of 10 started with Australia and ended with the Czech Republic: the UK was number 15, Germany number 17, and the US squeezed in at number 19.

This is what gets reported in the media, admiring the ever-happy Nordics and pointing out how America's (badly distributed) riches cannot buy happiness. What rarely gets reported is what actually goes into constructing these national scores: GDP per capita, social support (determined by asking if, when in trouble, people have relatives or friends to count on), healthy life expectancy

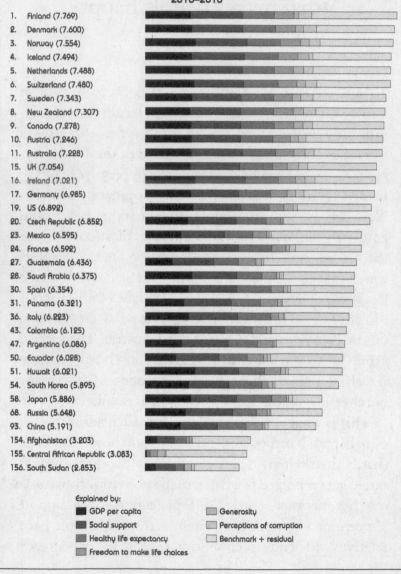

Happiness by country:
2016–2018

1.	Finland (7.769)
2.	Denmark (7.600)
3.	Norway (7.554)
4.	Iceland (7.494)
5.	Netherlands (7.488)
6.	Switzerland (7.480)
7.	Sweden (7.343)
8.	New Zealand (7.307)
9.	Canada (7.278)
10.	Austria (7.246)
11.	Australia (7.228)
15.	UK (7.054)
16.	Ireland (7.021)
17.	Germany (6.985)
19.	US (6.892)
20.	Czech Republic (6.852)
23.	Mexico (6.595)
24.	France (6.592)
27.	Guatemala (6.436)
28.	Saudi Arabia (6.375)
30.	Spain (6.354)
31.	Panama (6.321)
36.	Italy (6.223)
43.	Colombia (6.125)
47.	Argentina (6.086)
50.	Ecuador (6.028)
51.	Kuwait (6.021)
54.	South Korea (5.895)
58.	Japan (5.886)
68.	Russia (5.648)
93.	China (5.191)
154.	Afghanistan (3.203)
155.	Central African Republic (3.083)
156.	South Sudan (2.853)

Explained by:
- GDP per capita
- Social support
- Healthy life expectancy
- Freedom to make life choices
- Generosity
- Perceptions of corruption
- Benchmark + residual

(taken from the World Health Organization's assessment of 100 different health factors), freedom to make life choices (scored by answering the question "Are you satisfied or dissatisfied with your freedom to choose what you do with your life?"), generosity ("Have you donated money to a charity in the past month?"), and perceptions of corruption (throughout the government and within business).

As with all indices, this one contains a mix of components, including: a notoriously questionable indicator (national GDP converted to US dollars); answers that cannot be easily compared across cultures (perception of freedom to choose); and scores based on objective and revealing variables (healthy life expectancy). This mélange alone indicates that there should be a great deal of skepticism regarding any precise ranking—and this feeling is mightily reinforced when one looks closely at what never gets reported in the media: the actual country scores (accurate to the third decimal digit!). By coincidence, in 2019 I lectured in all of the world's three happiest countries—but, obviously, I was not able to notice that the Finns (7.769) are 2.2 percent happier than the Danes (7.600), who in turn are 0.6 percent happier than the Norwegians. The absurdity of all of this is obvious. Even ninth-place Canada has a combined score that is just 6.3 percent lower than Finland's. Given all the inherent uncertainties regarding the constituent variables and their simplistic, unweighted addition, would it not be more accurate, more honest (and, of course, less worthy of media attention), to at least round the scores to the nearest unit—or, better yet, do no individual ranking and just say which 10 or 20 countries make up the top group?

And then there is that remarkable lack of correlation between happiness and suicide—plotting the two variables for all European countries shows a complete absence of a relationship. Indeed, some of the happiest countries have relatively high suicide rates, and some rather unhappy places have a very low frequency of suicide.

But what, besides being Nordic and rich, makes people happy? Fascinating clues are provided by countries whose ranking appears out of place. That Afghanistan, the Central African Republic, and South Sudan are the three least happy countries of the 156 ranked nations is, sadly, quite expected (civil wars have been destroying them for far too long). But 23rd-place Mexico (a narco state with an extraordinarily high rate of violence and murder) ahead of France? Guatemala ahead of Saudi Arabia? Panama ahead of Italy? Colombia ahead of Kuwait? Argentina ahead of Japan? And Ecuador ahead of South Korea? Clearly, these pairs form a remarkable pattern: their second members are richer (often vastly so), more stable, less violent, and offer a considerably easier life than the first countries of every pair, whose commonalities are obvious—they may be relatively poor, troubled, and even violent, but they are all former Spanish colonies and hence overwhelmingly Catholic.

And all of them are in the top 50 (Ecuador is in the 50th spot), well ahead of Japan (58) and far ahead of China (93), the country that has been seen by naive Westerners as a veritable economic paradise full of happy shoppers. But while Louis Vuitton may be making a bundle in China, neither the massive malls nor the leadership of the all-knowing party make the Chinese happy; even the citizens of dysfunctional and much poorer Nigeria (85) feel happier.

The lessons are clear: if you cannot fit into the top 10 (not being Nordic, Dutch, Swiss, Kiwi, or Canadian), convert to Catholicism and start learning *castellano*. *¡Buena suerte con eso!*

The rise of megacities

Modernity means many things—rising affluence and mobility; inexpensive and instant communication; an abundance of affordable food; longer life expectancy—but an extraterrestrial observer sending periodic reconnaissance probes to Earth would be impressed by a shift easily observable from space: the increasing pace of urbanization as cities keep encroaching, amoeba-like, on the surrounding countryside, creating massive blobs of intense light throughout the night.

In 1800, less than 2 percent of the world's population lived in cities; by 1900 the share was still only about 5 percent. By 1950 it had reached 30 percent, and 2007 became the first year when more than half of humanity lived in cities. By 2016, the United Nations' comprehensive survey found 512 cities with a population greater than 1 million, with 45 of them larger than 5 million and 31 surpassing 10 million. This largest group has a special name: "megacities."

This continued concentration of humanity in ever-larger cities has been driven by advantages arising from the agglomeration of people, knowledge, and activities, often due to collocation of kindred companies: on the global level, think of London and New York, the financial capitals, and of Shenzhen in China's Guang-dong province, the capital of consumer electronics. Economies of scale bring many savings; interactions between producers, suppliers, and consumers are easier to handle; businesses have access to large pools of labor and diverse expertise; and (despite their crowding and environmental problems) the quality of life in large cit-ies attracts talent, now often from all over the world. Cities are places of countless synergies and investment opportunities, and they offer superior educations and rewarding careers. This is why many smaller cities—much like the surrounding countryside—are losing population, but megacities keep growing.

Ranking them by size is not straightforward, because assorted administrative boundaries yield different num-bers than when the megacities are considered as functional units. Tokyo, the world's largest megacity, has eight dif-ferent jurisdictional or statistical definitions, from the 23 wards of the old city, with fewer than 10 million people, to the National Capital Region area with nearly 45 million. The one used by the city administration is the Tokyo Major Metropolitan Region (*Tōkyō daitoshi-ken*), which is defined by commuting access within 70 kilometers of the city's massive twin-tower Metro-politan Government Building (*Tōkyō tochō*) in Shinjuku: the region now has some 39 million people.

The growth of megacities offers a perfect illustration of receding Western influence and the rise of Asia. In 1900, 9 of the world's 10 largest cities were in Europe

Megacities: 2018

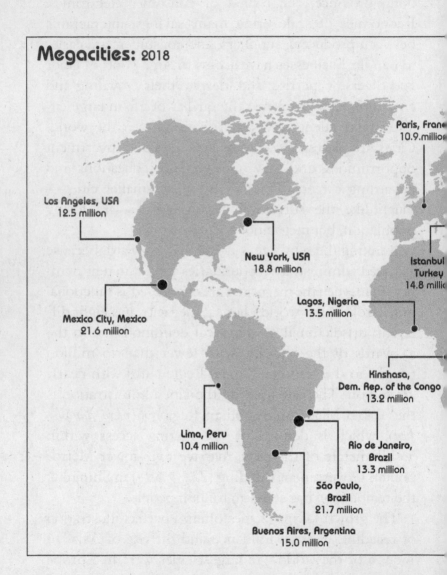

Paris, France
10.9 million

Los Angeles, USA
12.5 million

New York, USA
18.8 million

Istanbul
Turkey
14.8 million

Mexico City, Mexico
21.6 million

Lagos, Nigeria
13.5 million

Kinshasa,
Dem. Rep. of the Congo
13.2 million

Lima, Peru
10.4 million

Rio de Janeiro,
Brazil
13.3 million

São Paulo,
Brazil
21.7 million

Buenos Aires, Argentina
15.0 million

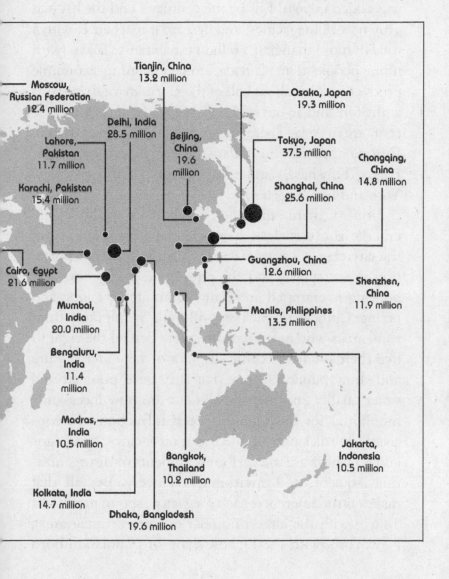

Tianjin, China
13.2 million

Osaka, Japan
19.3 million

Moscow,
Russian Federation
12.4 million

Delhi, India
28.5 million

Beijing,
China
19.6
million

Tokyo, Japan
37.5 million

Chongqing,
China
14.8 million

Lahore,
Pakistan
11.7 million

Shanghai, China
25.6 million

Karachi, Pakistan
15.4 million

Cairo, Egypt
21.6 million

Guangzhou, China
12.6 million

Shenzhen,
China
11.9 million

Mumbai,
India
20.0 million

Manila, Philippines
13.5 million

Bengaluru,
India
11.4
million

Madras,
India
10.5 million

Jakarta,
Indonesia
10.5 million

Bangkok,
Thailand
10.2 million

Kolkata, India
14.7 million

Dhaka, Bangladesh
19.6 million

and the United States. In 1950 New York and Tokyo were the only megacities, and the third, Mexico City, was added in 1975. But by the century's end the list had grown to 18 megacities, and by 2020 it reached 35 with a total of more than half a billion inhabitants. Tokyo (with more people than Canada, and generating economic product equal to about half of the German total) remains at the top, and 20 out of the 35 megacities (nearly 60 percent) are in Asia. There are six in Latin America, two in Europe (Moscow and Paris), three in Africa (Cairo, Lagos, Kinshasa), and two in North America (New York and Los Angeles).

None of them ranks high on all major quality-of-life criteria: Tokyo is clean, its residential areas not far from the city center are remarkably quiet, public transportation is exemplary, and the crime rate is very low; but housing is cramped and daily commutes are long and taxing. Chinese megacities—all built by migrants from rural areas who (until recently) were denied the right to live there—have become displays of new architecture and shiny public projects, but they have poor air and water quality and their inhabitants are now incessantly monitored for the slightest social infractions. In contrast, few rules prevail in African megacities, and Lagos and Kinshasa are the very embodiments of disorganization, squalor, and environmental decay. But all that makes little difference; each megacity—no matter if it is Tokyo (with the largest number of starred restaurants), New York (with the highest share of population born abroad), or Rio de Janeiro (with a murder rate approaching 40 per 100,000)—continues to attract people. And the United Nations has forecast the emergence of 10 additional megacities by 2030: six in Asia (including India's Ahmedabad and Hyderabad), three in Africa (Johannesburg, Dar es Salaam, Luanda), and Colombia's Bogotá.

COUNTRIES

Nations in the Age of Globalization

The First World War's
extended tragedies

Few recent 100-year anniversaries resonated so loudly as November 2018's marking of the end of the world's first truly global armed conflict. The war's enormous carnage scarred the memory of a generation, but its most tragic legacy was the resulting Communist rule in Russia (1917), Fascist rule in Italy (1922), and Nazi rule in Germany (1933). Those developments led to the Second

The Battle of the Somme, 1916: British troops and Mark I tank

World War, which killed even more people and left direct and indirect legacies—including NATO vs. Russia, and a divided Korea—that still trouble our lives.

Even though the Second World War was deadlier, a case can be made that the first war constituted the critical disaster, as it gave rise to so much of what followed. True, the second war deployed far greater advances in destructive power, including the fastest fighter planes ever powered by reciprocating engines; enormous four-engine bombers (the B-17); missiles (the German V-1 and V-2); and, at the war's very end, the nuclear bombs that destroyed Hiroshima and Nagasaki.

In comparison, the First World War, with its entrenched and barely shifting fronts, was a decidedly less dynamic conflict. But a closer look shows that purely technical advances were indeed critical for lengthening its duration and adding to its death toll.

Leaving aside the use of poisonous gases in combat (never repeated on such a scale), several key modes of modern warfare were developed and even perfected during the earlier conflict. The first diesel-powered submarines were used on long forays to attack convoys of merchant ships. The first tanks were deployed in combat. The first bombing raids, using both dirigibles and airplanes, were mounted. The first battle-ready aircraft carrier was launched in 1914. The French successfully tested portable transmitters enabling voice communication from the air to the ground in 1916, and from air to air in 1917, beginning the long road toward ever-smaller, ever-more-usable electronic components.

But amid all these developments, we must single out the momentous innovation that allowed a blockaded Germany to endure its two-front war for four years: the synthesis of ammonia. When the war began, the British navy cut Germany off from imports of Chilean nitrates needed to produce explosives. But thanks to a remarkable coincidence, Germany could instead supply itself with homemade nitrates. In 1909, Fritz Haber, a professor at the University of Karlsruhe, had ended the long-running quest for synthesizing ammonia from its elements. Nitrogen and hydrogen were combined under high pressure and in the presence of a catalyst, to make ammonia (NH_3).

By October 1913, BASF—then the world's leading chemical conglomerate, under the leadership of Carl Bosch—had commercialized the process at the world's first ammonia plant, in Oppau in Germany. This synthetic ammonia was to be used in the production of such solid fertilizers as sodium or ammonium nitrate (see THE WORLD WITHOUT SYNTHETIC AMMONIA, p. 221).

But the war began less than a year later, and instead of converting ammonia into fertilizer, BASF began to mass-produce the compound for conversion into nitric acid to be used in the synthesis of wartime explosives. A larger ammonia plant was completed by April 1917 in Leuna, west of Leipzig, and the combined production of the two plants sufficed to support Germany's manufacture of explosives until the war's end.

The new capability of industry to find ways around every shortage helped to drag out the First World War, adding millions of casualties. This terrifyingly modern development belies the war's primitive image, framed as it so often is by prolonged stalemates in muddy trenches—and it paved the way to even greater carnage a generation later.

Is the US really exceptional?

Belief in "American exceptionalism"—that unique blend of ideals, ideas, and love of liberty made so powerful by great technical and economic accomplishments—is alive and well. Even former President Obama, the man known for his unemotional approach to governing and hence a reluctant endorser to begin with, has come around. Early in his presidency (in April 2009), he affirmed his belief by essentially denying it: "I believe in American exceptionalism, just as I suspect that the Brits believe in British exceptionalism and the Greeks believe in Greek exceptionalism." By May of 2014, he had relented: "I believe in American exceptionalism with every fiber of my being."

But such proclamations mean nothing if they cannot stand up to the facts. And here, what really matters is not the size of a country's gross domestic product or the number of warheads or patents it may possess, but the variables that truly capture its citizens' physical and intellectual well-being. These variables are simply life, death, and knowledge.

Infant mortality is an excellent proxy for a wide range of conditions, including income, quality of housing, nutrition, education, and investment in healthcare. Very few babies die in those affluent countries where people live in

LIFE EXPECTANCY
Years

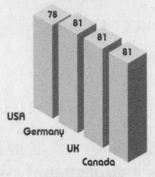

USA	Germany
28th	**24th**
UK	Canada
22nd	**13th**

(36 OECD countries)

INFANT MORTALITY
Deaths per 1,000 live births

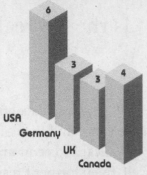

USA	Germany
33rd	**19th**
UK	Canada
24th	**30th**

(36 OECD countries)

OBESITY
% of population

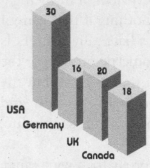

USA	Germany
1st	**19th**
UK	Canada
3rd	**4th**

(36 OECD countries)

HAPPINESS
Highest 7.769, lowest 2.853

USA	Germany
19th	**17th**
UK	Canada
15th	**9th**

(156 countries)

good housing and where well-educated parents (themselves well nourished) feed their children properly and have access to medical care (see THE BEST INDICATOR OF QUALITY OF LIFE? TRY INFANT MORTALITY, p. 8). So how does the United States rank among the world's roughly 200 nations? The latest available comparison shows that with 6 of every 1,000 live-born babies dying in the first year of life, the US does not figure among the top 25 nations. Its infant mortality is far higher than in France (4), Germany (3), and Japan (2). And the US rate was 50 percent higher than in Greece (4), a country portrayed in the press as an utter basket case ever since the financial crisis.

Excusing that very poor rating by saying that the European countries have homogeneous populations does not work: modern France and Germany are full of recent immigrants (just spend some time in Marseille or Düsseldorf). What matters more is parental knowledge, good nutrition, the extent of economic inequality, and access to universal healthcare, the United States being (notoriously) the only modern affluent country without the latter.

And looking at the journey's end gives an almost identically poor result: recent US life expectancy (nearly 79 years for both sexes) does not even rank among the top two dozen countries worldwide, and is—again—behind Greece (about 81), as well as South Korea (nearly 83). Canadians live more than three years longer on average and Japanese (about 84) nearly six years longer as compared to their US counterparts.

Educational achievements of US students (or a lack thereof) are scrutinized with every new edition of the Organisation for Economic Co-operation and Development's Program for International Student Assessment, or PISA. The latest results (2018) for 15-year-olds show that, in math, the United States ranks just below Russia, Slovakia, and Spain, but far lower than Canada, Germany, and Japan. In science, US schoolchildren place just below the mean PISA score (497 versus 501); in reading, they are barely above it (498 versus 496)—and they are far behind all the populous, affluent Western nations. PISA, like any such study, has its weaknesses, but large differences in relative rankings are clear: there is not even a remote indication of any exceptional US educational achievements.

American readers might find these facts discomforting, but there is nothing arguable about them. In the United States, babies are more likely to die and high schoolers are less likely to learn than their counterparts in other affluent countries. Politicians may look far and wide for evidence of American exceptionalism, but they won't find it in the numbers, where it matters.

Why Europe should be more pleased with itself

On January 1, 1958, Belgium, France, Italy, Luxembourg, the Netherlands, and the Federal Republic of Germany jointly formed the European Economic Community (EEC) with the aim of economic integration and free trade within a customs union.

Although the immediate goals were explicitly economic, the EEC's aspirations were always far higher. In the founding document, the Treaty of Rome, the member states declared their determination "to lay the foundations of an ever-closer union among the peoples of Europe" and "to ensure the economic and social progress of their countries by common action to eliminate the barriers which divide Europe." At that time, these goals seemed to be quite unrealistic: Europe was divided not only by national prejudices and economic inequalities but also, most fundamentally, by the Iron Curtain, which ran from the Baltic to the Black Sea, with Moscow controlling the nations to the east of it.

That Soviet control was reasserted after the failure of the Prague Spring in 1968 (as the attempted Czechoslovak reforms ended in the Soviet invasion of the country), while the EEC continued to accept new members: the United Kingdom, Ireland, and Denmark in 1973; Greece

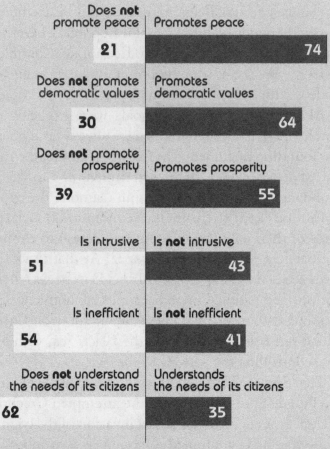

Percentage of Europeans who say the European Union...

Does **not** promote peace	Promotes peace
21	74

Does **not** promote democratic values	Promotes democratic values
30	64

Does **not** promote prosperity	Promotes prosperity
39	55

Is intrusive	Is **not** intrusive
51	43

Is inefficient	Is **not** inefficient
54	41

Does **not** understand the needs of its citizens	Understands the needs of its citizens
62	35

Note: Percentages are medians based on 10 European countries.

in 1981; Spain and Portugal in 1986. And then, after the USSR collapsed in 1991, the way was open to pan-European integration. In 1993, the Maastricht Treaty established the European Union; in 1999, a common currency, the euro, was created; and 27 nations now belong to the Union.

The EU has just over 450 million people, less than 6 percent of the global population, but it generates nearly 20 percent of the world's economic output, as against almost 25 percent for the United States. It accounts for nearly 15 percent of global exports of goods—a third more than the United States—including cars, airliners, pharmaceuticals, and luxury goods. Moreover, half of its 27 members are among the top 30 countries in terms of quality of life, as measured by the United Nations' Human Development Index.

And yet, today, the EU is witnessing mounting worries and disaffection. The bonds of union are loosening, and the UK has left outright.

Within Europe, the commentariat offers endless explanations of this new centrifugal temper: the excessive bureaucratic control exercised from Brussels; the reassertion of national sovereignty; and poor economic and political choices, notably the adoption of a common currency without common fiscal responsibility.

I must confess that I am puzzled. As somebody who was born during Nazi occupation, who grew up on the wrong side of the Iron Curtain, and whose family history is typical of Europe's often-complicated national and linguistic origins, I see today's Europe—shortcomings and all—as a staggering outcome, too good to be believed. Surely these achievements are worthy of redoubled efforts at compromises to reunite it.

Instead, decades of peace and prosperity have been taken for granted, and lapses and difficulties (some inevitable, some unpardonable) have served to reignite old biases and animosities. My wish for Europe: make it work. The failure to do so cannot be contemplated lightly.

Brexit: Realities that matter
most will not change

What will really be different in the post-Brexit UK? Of course, much has already changed during the unexpectedly protracted run-up to the event, and what has taken place can be best described by using the words the English language gained thanks to the last successful invasion of those isles: the country has been through a disorienting period of accusations, acrimony, condemnations, delusions, falsifications, illusions, recriminations, and tested civility.

But what will really change five or ten years down the road, as far as the fundamental determinants of the nation's life are concerned? First things first. We all have to eat, and modern societies have been extraordinarily successful in supplying an unprecedented variety of foodstuffs at a generally affordable cost. We have to energize our buildings, our industries, and our transport by incessant flows of fuel and electricity. We have to produce—and renew—the material foundations of our societies by manufacturing, building, and maintenance. And we need adequate infrastructures (schools, health, and elderly care) to educate people and to care for them in sickness and old age. Everything else is secondary.

On all of these scores the accounts are clear. The UK

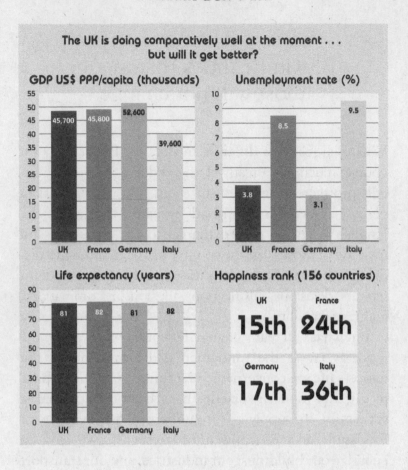

The UK is doing comparatively well at the moment . . .
but will it get better?

GDP US$ PPP/capita (thousands)

UK 45,700 · France 45,800 · Germany 52,600 · Italy 39,600

Unemployment rate (%)

UK 3.8 · France 8.5 · Germany 3.1 · Italy 9.5

Life expectancy (years)

UK 81 · France 82 · Germany 81 · Italy 82

Happiness rank (156 countries)

| UK | France |
| 15th | 24th |

| Germany | Italy |
| 17th | 36th |

has not been self-sufficient in food production for a few centuries, and its dependency on imports has doubled from about 20 percent in the early 1980s to 40 percent in recent years, and in the short term nothing short of draconian food rationing (and no fresh produce in winter) can significantly reduce this import dependence. Three-quarters of British food imports come from the EU, but

Spanish vegetable growers and Danish bacon producers will remain as eager to ship their products as the British consumers will be to buy them, and hence there will be no demand-destroying taxes or pricing.

The last time the UK was a net exporter of energy (oil and gas from the North Sea) was in 2003, and in recent years the country has been importing 30–40 percent of its primary energy—natural gas above all. Again, no major shifts will take place in the near future, and the well-supplied global energy market will assure the continuation of affordable import prices.

The UK—once the unrivaled inventor and pioneer of modern science-based manufacturing (it is the country of Michael Faraday, Isambard Kingdom Brunel, James Clerk Maxwell and Charles Algernon Parsons, after all)—is already more deindustrialized than Canada, historically the least industrialized Western nation. In 2018, manufacturing accounted for 9 percent of the British GDP, compared to 10 percent in Canada, 11 percent in the US, and, respectively, 19, 21, and 27 percent in such remaining manufacturing superpowers as Japan, Germany, and South Korea . . . and 32 percent in Ireland, whose share now beats even China's 29 percent. Yet, again, no overnight switch in political arrangements can turn this historic tide.

Much as in the rest of Europe, the UK's modern education provision has put excessive stress on quantity over quality, its healthcare system labors under many much-studied constraints (easily illustrated by a stream of reports about overwhelmed National Health Service employees and overburdened hospitals), and its aging population will require more resources. The country's old-age dependency ratio (number of people 65 and older as a share of all economically active people 20–64 years of age), standing at 32 percent in 2020—still slightly lower than in France or Germany—will increase to 47 percent by the year 2050. No governmental intervention and no declaration of regained sovereignty and severance from the bureaucrats of Brussels will have any effect on this inexorable process.

Given these fundamental realities, a rational observer must wonder what tangible differences, what clear benefits could any reassertion of British insularity bring. False claims can be painted on buses, extravagant promises are easy to make, feelings of pride or satisfaction may become fleetingly convincing—but none of those intangibles can change what the UK has become: an aging nation; a deindustrialized and worn-out country whose per capita GDP is now just over half of the Irish mean (something that Swift, Gladstone, or Churchill would find utterly unfathomable); another has-been power whose claim to uniqueness rests on having too many troubled princes and on exporting costumed TV series set in fading country mansions staffed with too many servants.

Concerns about Japan's future

On September 2, 1945, representatives of the Japanese government signed the instrument of surrender on the deck of the USS *Missouri*, anchored in Tokyo Bay. So ended perhaps the most reckless of all modern wars, the outcome of which was decided by US technical superiority even before it started. Japan had already lost in

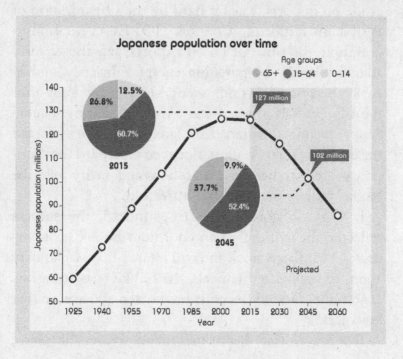

material terms when it attacked Pearl Harbor—in 1940, the United States produced roughly 10 times as much steel as Japan did, and during the war the difference grew further.

The devastated Japanese economy did not surpass its prewar peak until 1953. But, by then, the foundations had been laid for the country's spectacular rise. Soon its fast-selling exports ranged from the first transistor radios (Sony) to the first giant crude oil tankers (Sumitomo). The first Honda Civic arrived in the United States in 1973, and by 1980, Japanese cars claimed 30 percent of the US market. In 1973–74, Japan, totally dependent on crude oil imports, was hit hard by the Organization of Petroleum Exporting Countries (OPEC)'s decision to quintuple the price of its oil exports, but the country adjusted rapidly by pursuing energy efficiency, and in 1978 it became the world's second-largest economy after the US. By 1985, the yen was so strong that the United States, feeling threatened by Japanese imports, forced the currency's devaluation. But even afterward the economy soared: in the five years following January 1985, the Nikkei index rose more than threefold.

It was too good to be true; indeed, the success reflected the working of an enormous bubble economy driven by inflated stock and real estate prices. In January 2000, 10 years after its peak, the Nikkei was at half its 1990 value, and only recently has it risen above even that low mark.

Iconic consumer electronics manufacturers like Sony,

Toshiba, and Hitachi now struggle to be profitable. Toyota and Honda, global automotive brands once known for their unmatched reliability, are recalling millions of vehicles. Since 2014, Takata's defective airbags have resulted in the biggest recall of a manufactured part ever. In 2013, GS Yuasa's unreliable lithium-ion batteries caused problems for the new Boeing 787. Add to this frequently changing governments, the March 2011 tsunami followed by the Fukushima disaster, constant concerns about an unpredictable North Korea, and worsening relations with China and South Korea, and you get a worrisome picture indeed.

And there is an even more fundamental problem. In the long run, the fortunes of nations are determined by population trends. Japan is not only the world's fastest-aging major economy (already every fourth person is older than 65, and by 2050 that share will be nearly 40 percent), its population is also declining. Today's 127 million will shrink to 97 million by 2050, and forecasts show shortages of the young labor force needed in construction and healthcare. Who will maintain Japan's extensive and admirably efficient transportation infrastructures? Who will take care of the millions of old people? By 2050, people above the age of 80 will outnumber the children.

Fortunes of all major nations have followed specific trajectories of rise and retreat, but perhaps the greatest difference in their paths has been the time they spent at the top of their performance: some had a relatively prolonged plateau followed by steady decline (both the British empire and the 20th-century United States fit this pattern); others had a swift rise to a brief peak, followed by a more or less rapid decline. Japan is clearly in the latter category. Its swift post–Second World War ascent ended in the late 1980s, and it's been downhill ever since: in a single lifetime, from misery to an admired—and feared—economic superpower, then on to the stagnation and retreat of an aging society. The Japanese government has been trying to find some way out—but radical reforms are not easy in a gerrymandered country that still cannot seriously contemplate even moderate-scale immigration and that is yet to make real peace with its neighbors.

How far can China go?

Some milestones are anticipated for years. How many articles have been written on how China will surpass the United States to become the world's largest economy by—take your pick—2015, 2020, or 2025? The timing depends on what monies we use. It's already happened in terms of purchasing power parity (PPP), which compares the economic product of different countries by eliminating distortions caused by fluctuations in the

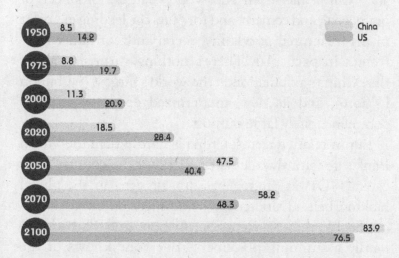

Old-age dependency ratio, China–US

	China	US
1950	8.5	14.2
1975	8.8	19.7
2000	11.3	20.9
2020	18.5	28.4
2050	47.5	40.4
2070	58.2	48.3
2100	83.9	76.5

exchange rates of their national currencies. According to the International Monetary Fund (IMF), in 2019 China's PPP-adjusted GDP was about 32 percent ahead of the US total.

If you rely instead on the yuan-to-US-dollar exchange rate, the United States is well ahead: about 50 percent higher in 2019 ($21.4 trillion versus $14.1 trillion). But even with the recent slowdown in Chinese GDP growth—from double digits to an official rate of between 6 and 7 percent a year, and, in reality, less than that—it is still considerably higher than growth in the United States. It is thus only a matter of time before China becomes No. 1, even in nominal terms.

The path to No. 1 status began in 1978, when the country embraced economic modernization, leaving behind three decades of gross mismanagement. For decades China has been the world's largest producer of grain, coal, and cement, and for years the leading exporter of manufactured goods in general and consumer electronics in particular. There's nothing surprising about this: China's population is the world's largest (1.4 billion in 2016), and its new, modernized economy requires commensurately large outputs.

But in relative terms, China is hardly rich: the World Bank's generously calculated PPP put the country's per capita GDP at $19,504 in 2019, or 73rd in the global ranking, behind Montenegro and Argentina and just ahead of the Dominican Republic, Gabon, and Barbados— hardly a stunning placement. Everyone knows of the rich Chinese who buy real estate in Vancouver and London and diamond-encrusted watches at Galeries Lafayette in Paris, but they constitute a tiny minority.

The GDP and the number of *nouveaux riches* are misleading measures of the actual quality of life in China. The environment has kept on deteriorating. Air pollution in the country's cities is incredibly bad: according to the World Health Organization, the maximum acceptable level of particulates with diameters under 2.5 nanometers is 25 micrograms per cubic meter of air, but many Chinese cities have repeatedly exceeded 500 µg/m³. Some cities have even seen maximums above 1,000. In 2015, Beijing averaged 80 µg/m³, compared with less than 10 for New York City. Such extremely high levels of pollution increase the incidence of respiratory and heart diseases, and shorten expected lifespans.

Water pollution is also endemic. Nearly half of those living in rural areas in China lack modern sanitation. The country has less arable land per capita than India, and unlike the much smaller Japan, it could never rely largely on imports. China's oil and natural gas resources are inferior to the US endowment, with recent crude oil imports accounting for more than 60 percent of total consumption, while the US is now only a minor importer. And it is better not to think about a Fukushima-like disaster in a country where so many new nuclear reactors have quickly been built in densely populated coastal provinces, or about another pandemic starting in one of the popular wet markets.

Finally, the country's population is aging rather rapidly—that's why the Communist Party abandoned its one-child-per-couple policy in 2015—and, as a result, its demographic advantage is already receding. The ratio of economically active to economically dependent people peaked in 2010, and as the ratio declines, so will China's industrial dynamism.

We've seen it all before. Compare the Japan of 1990, whose rise appeared to challenge the entire Western world, with the Japan of 2020, after 30 years of economic stagnation (see CONCERNS ABOUT JAPAN'S FUTURE, p. 69). This is perhaps the best insight into the likely contrast between the China of 2020 and that of 2050.

India vs. China

India as No. 1? It's on the cards: India will soon supplant China as the world's most populous country. The question is whether India will also rise to challenge China as an economic powerhouse.

At least since the unraveling of the Roman empire, successive Chinese dynasties have ruled over more people than any other government. China had about 428 million people in 1912, when imperial rule ended; 542 million in 1949, when the Communists took power; 1.27 billion by the year 2000; and about 1.4 billion by the end of 2019. The slowing growth rate is the direct result of the one-child policy, adopted in 1979 and ended in 2015 (see the previous chapter). Meanwhile, India's population expanded from 356 million in 1950 to 1.05 billion in 2000, and to 1.37 billion by the end of 2019.

China's edge has been shrinking fast—and, given the reliability of short-term demographic forecasts, it seems clear that India will surpass China's total no later than 2025 (according to the latest UN median forecast), and perhaps as early as 2023.

Meanwhile, it is fascinating to compare the two giga-states. Both countries selectively abort many girls, creating abnormal sex ratios at birth. The normal ratio is

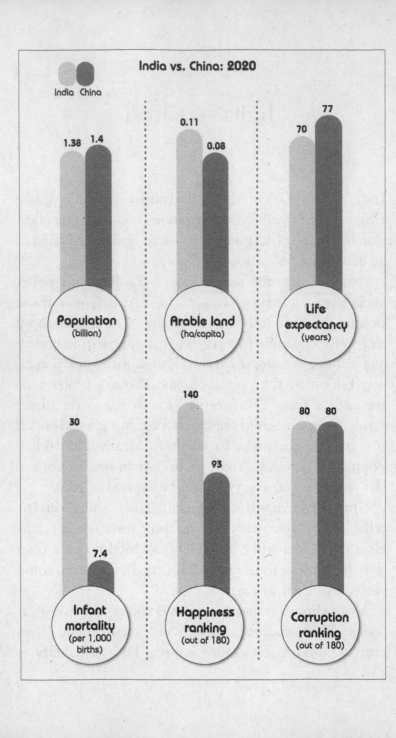

1.06 males per one female, but India stands at 1.12 and China at 1.15.

Both countries are riddled with corruption: the latest Corruption Perceptions Index, from Transparency International, puts India and China at 80th among the 180 included countries (Denmark is the least corrupt and Somalia the most). In both countries, economic inequality as measured by the Gini index is very high— about 48 in India and 51 in China (compared to 25 in Denmark, 33 in the UK, and 38 in the US). And in both countries the moneyed classes compete in ostentatious consumption, collecting expensive cars and palatial residences. Mukesh Ambani, chairman of Reliance Industries Limited, owns the world's most expensive private residence; his Antilia, a 27-story skyscraper completed in 2012, has a perfect view of Mumbai's slums.

But there are also fundamental differences. Rapid economic growth since 1980 has made China by far the richer of the two, with a nominal GDP (per the IMF estimate for 2019) nearly five times that of India ($14.1 trillion versus $2.9 trillion). In 2019, China's per capita average, measured in terms of purchasing power parity, was (according to the IMF) more than twice as high as India's ($20,980 versus $9,030).

On the other hand, China is a tightly controlled one-party state run by a politburo of seven aged men, while India continues as a highly imperfect but undeniably democratic polity. In 2019, Freedom House assigned India 75 points on its freedom index, compared with a measly 11 points for China (the UK got 93 and Canada 99).

Another comparison is equally revealing: one of China's top high-tech achievements is employing ferocious censorship of the Internet and highly intrusive monitoring as part of the new, pervasive Social Credit System; one of India's great high-tech achievements is its disproportionate contribution to high-tech corporate leadership at home and abroad. Many Indian emigrants have risen to leadership in Silicon Valley: Sundar Pichai at Google, Satya Nadella at Microsoft, Shantanu Narayen at Adobe, and Sanjay Jha, the former CEO of Global-Foundries, to name the most prominent ones.

It will be fascinating to see to what extent India can replicate China's economic success. And China, for its part, must cope with its loss of the demographic dividend: since 2012, its dependency ratio—the number of those who are too young or too old to work divided by the number of people of working age—has been rising (it is now just over 40 percent). The question is whether the country will become old before it can become truly rich. Both countries have enormous environmental problems and both will be challenged to feed their populations—but India has about 50 percent more farmland.

One last complication: these two nuclear powers have yet to sign a binding treaty to end their territorial disagreement in the Himalayas. They've come to blows over the matter, most notably in 1962. Things can get touchy when rising powers sit athwart a disputed border.

And yet this conflict is not India's greatest immediate challenge. More pressing are the need to further lower its fertility rate as rapidly as possible (everything else being equal, that raises per capita income), the challenges of maintaining basic food self-sufficiency (a country of more than 1.4 billion is too large to rely on imports), and finding a way out of the deteriorating relations between the country's Hindus and Muslims.

Why manufacturing remains important

Manufacturing has become both bigger and smaller. Between 2000 and 2017, the worldwide value of manufactured products has more than doubled, from $6.1 trillion to $13.2 trillion. Meanwhile, the *relative* importance of manufacturing is dropping fast, retracing the earlier retreat of agriculture (now just 4 percent of the world's economic product). Based on the United Nations' uniform national statistics, the manufacturing sector's contribution to global economic product declined from 25 percent in 1970 to less than 16 percent by 2017.

The decline has registered in the stock market, which values many service companies above the largest manufacturing firms. At the end of 2019, Facebook—that purveyor of constant selfies—had a market capitalization of almost $575 billion, nearly three times more than Toyota, the world's premier maker of passenger cars. And SAP, Europe's largest software provider, was worth about 60 percent more than Airbus, Europe's largest maker of jetliners.

And yet manufacturing is still important for the health of a country's economy, because no other sector can generate nearly as many well-paying jobs. Take Facebook, which at the end of 2019 had about 43,000 employees

Manufacturing creates jobs: only 2 out of the top 10 manufacturing countries have recent unemployment above 5%

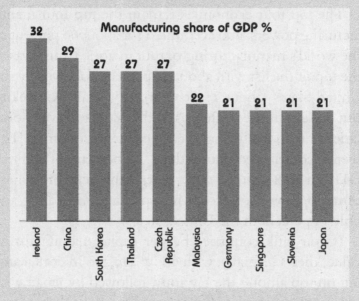

Manufacturing share of GDP %

Ireland	China	South Korea	Thailand	Czech Republic	Malaysia	Germany	Singapore	Slovenia	Japan
32	29	27	27	27	22	21	21	21	21

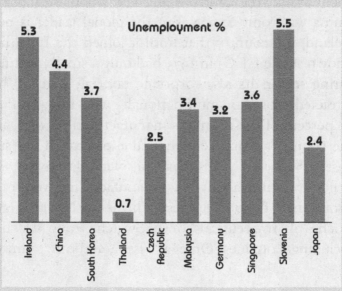

Unemployment %

Ireland	China	South Korea	Thailand	Czech Republic	Malaysia	Germany	Singapore	Slovenia	Japan
5.3	4.4	3.7	0.7	2.5	3.4	3.2	3.6	5.5	2.4

versus the 370,000 or so that Toyota had during the 2019 fiscal year. Making things still matters.

The top four economies remain the top four manufacturing powers, and accounted for about 60 percent of the world's manufacturing output in 2018. China was at the top of the list with about 30 percent, followed by the United States (about 17 percent), Japan, and Germany. But these countries differ markedly in the relative importance of manufacturing to each of their economies. The sector contributed more than 29 percent of China's GDP in 2018—in the same year, manufacturing's share came to about 21 percent in Japan and Germany, and only 12 percent in the United States.

If you rank countries by per capita manufacturing value, then Germany, with about $10,200 in 2018, came out on top among the big four, followed by Japan with about $7,900, the United States with about $6,800, and China with only $2,900. But the global leader is now Ireland, the country that until it joined the EU (then known as the EEC) in 1973 had only a small manufacturing sector. Its low corporate tax (12.5 percent) has attracted scores of multinationals, who now produce 90 percent of the country's manufactured exports, and the country's manufacturing value per capita has surpassed $25,000 a year, ahead of Switzerland's $15,000. When you think about Swiss manufacturing, you think about such famous domestic firms as Novartis and Roche (pharmaceuticals) or the Swatch Group (watches, including Longines, Omega, Tissot, and other famous brands). When you think of Irish manufacturing, you think about Apple, Johnson & Johnson, or Pfizer—foreigners all.

Countries where manufactured goods account for more than 90 percent of total merchandise trade include not only China and Ireland, but also Bangladesh, the Czech Republic, Israel, and South Korea. Germany is close to 90 percent; the US share is below 70 percent.

The net balance of international trade in manufactured items is also revealing because it indicates two things: the extent to which a nation can satisfy its own need for products; and the demand for its products abroad. As expected, Switzerland, Germany, and South Korea have large surpluses, while the United States had another record goods trade deficit in 2018 at $891 billion, or some $2,700 per capita—the price to pay for importing electronics, clothes, shoes, furniture, and kitchen gadgets from Asia.

But the United States enjoyed generations of manufacturing trade surpluses until 1982; China had chronic deficits until 1989. What are the chances of the United States redressing its massive manufactured trade imbalance with China, or of India replicating China's manufacturing success?

Russia and the USA:
How things never change

Tensions between Russia and the USA that arose during the second decade of the 21st century are just the latest reincarnation of the long-running superpower rivalry. In August 2019, the United States withdrew from the Intermediate-Range Nuclear Forces Treaty with Russia; both sides are developing new missiles; and the

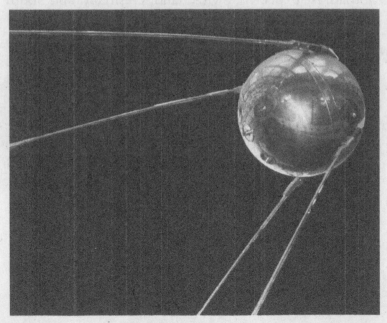

The Sputnik

countries have been sparring over the future of the once-Soviet Ukraine.

When looking back, it is clear that one of the decisive moments in their decades-long confrontation took place on Friday, October 4, 1957, when the Soviet Union launched Sputnik 1, the first artificial satellite. Technically, it was a modest affair: a sphere 58 centimeters in diameter, weighing almost 84 kilograms and sprouting four rodlike aerials. Although its three silver-zinc batteries made up some 60 percent of the total mass, they rated only 1 watt—good enough to broadcast rapid shrill beeps at 20.007 and 40.002 megahertz for three weeks. The satellite circled the planet 1,440 times before plunging to a fiery death on January 4, 1958.

Sputnik should have come as no surprise. Both the Soviets and the United States had revealed their intent to put satellites in orbit during the International Geophysical Year (1957–58), and the Soviets had even published some technical details before the launch. But that was not how the public perceived the little beeping sphere in late 1957.

The Western world reacted with awe; the United States with embarrassment. And the embarrassment only deepened in December, when the Vanguard TV3 rocket, its launch hastily scheduled to counter the Sputnik effect, blew up on the launching pad at Cape Canaveral just two seconds after liftoff. Members of the Soviet delegation to the United Nations asked their US counterparts whether they would like to receive technical assistance under the Soviet program for undeveloped countries.

This public humbling led to calls for accelerating the country's space program, for erasing the perceived technical lag, and for boosting education in mathematics and science. The shock that the US school system received was perhaps the greatest in its history.

All this had great personal significance for me. In October 1957 I was a teenager in Czechoslovakia, and every day, when I walked to school, I looked into West Germany, inaccessible behind barbed wire and minefields. It might as well have been a different planet. Soviet premier Nikita Khrushchev had recently told the West, "We will bury you," and now his boasts about the supremacy of Communist science and engineering were finding support in the United States' near-panicky reactions. This latest demonstration of Soviet power led many of us to fear that it would not come to an end in time for our generation.

But it turned out that there had never been a real scientific or engineering gap: the United States soon gained decisive primacy in launching satellites for communications, weather forecasting, and espionage. Less than a dozen years after the surprise of Sputnik, Neil Armstrong and Buzz Aldrin stood on the Moon—a place no Soviet cosmonaut would ever reach.

And 11 years after Sputnik, the Soviet empire did weaken—if only temporarily—during the Prague Spring, when Czechoslovakia tried to adopt a freer form of its (still Communist) rule. As a result, even Czechs who were not members of the Communist Party could get passports to travel to the West. So, in August 1969, my wife and I landed in New York, just weeks before the borders were closed for another two decades.

In 1975, shortly after we moved from the United States to Canada, the first major exhibition at the newly completed Winnipeg Convention Center showcased the Soviet space program. A full-scale Sputnik model was hung by wires above the main lobby. As I rode the escalator and looked up at that shiny sphere, I was transported back to October 4, 1957, when for me the beeping satellite signaled not the glories of engineering and science, but the fear that Soviet power would go on for the rest of my life.

We made it out, but as the French say, *plus ça change, plus c'est la même chose.*

Receding empires:
Nothing new under the sun

Keeping an empire, be it a real one (with an emperor or an empress) or a de facto one (defined by economic and military might and sustained by the projection of power and by shifting alliances), has never been easy. Comparing the longevity of empires is difficult, due to their different degrees of centralization and actual exercise of effective territorial, political, and economic control. But one finding stands out: despite the increasing military, technical, and economic capabilities of major nations, keeping large empires for extended periods of time has become more difficult.

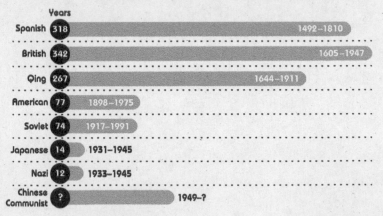

Longevity of recent empires and "empires"

	Years	
Spanish	318	1492–1810
British	342	1605–1947
Qing	267	1644–1911
American	77	1898–1975
Soviet	74	1917–1991
Japanese	14	1931–1945
Nazi	12	1933–1945
Chinese Communist	?	1949–?

When in 2011 Samuel Arbesman, at that time at the Institute for Quantitative Social Science at Harvard University, analyzed the lifespans of 41 ancient empires that existed between 3000 BCE and 600 CE, he found that their mean duration was 220 years, but that the distribution of imperial lifespans was highly skewed, with those empires enduring at least 200 years being roughly six times as common as those surviving for eight centuries. Moreover, the three most durable empires—Mesopotamian Elam, lasting ten centuries; and Egypt's Old and New Kingdoms, each surviving for five centuries—reached their maturity before 1000 BCE (Elam about 1600 BCE; the Egyptian kingdoms, 2800 and 1500 BCE).

There has been no shortage of empires post 600 CE, but a closer look reveals no gains in longevity. Of course, China continued to have some form of imperial rule until 1911, but that included a dozen different dynasties—including those set up by foreign invaders, the short-lived Mongolian Yuan (1279–1368) and the Manchu Qing (1644–1911)—that exercised various degrees of control over shrinking and expanding territories, often with tenuous claims on the northern and western regions beyond the Han core.

The timing of the Spanish and British empires is highly arguable. Taking 1492 as the beginning of the Spanish empire and 1810 as its de facto end, means just over three centuries of rule from Madrid (or, after 1584, from El Escorial). And should we time the British empire since 1497 (John Cabot's voyage to North America) or 1604 (the Treaty of London, concluding the Anglo-Spanish War)—and its end (leaving aside the remaining micro-possessions of overseas territories, ranging from Anguilla to Turks and Caicos) as being in 1947 (the loss of India) or 1960 (when Nigeria, Africa's most populous nation, became independent)? The later dates would give us 356 years.

And there was no empire able to last throughout the entire 20th century. The last Chinese dynasty, Qing, ended in 1911 after 267 years of rule, and the new Communist empire was set up only in 1949. The Soviet empire, the successor of the Romanovs, eventually regained control of most of the territory formerly ruled by the tsars (Finland and parts of Poland being the major exceptions), and after the Second World War it extended its control over the countries of eastern and central Europe as the Iron Curtain descended from the Baltic to the Black Sea.

During the Cold War years, the empire looked mighty to NATO planners and Washington policymakers, but from the inside (I lived under it until my 26th year) it looked less formidable. Even so, it was a surprise that, eventually, it dissolved so easily; it lasted from the first week of November 1917 to the last week of December 1991—74 years and a month, an average European male lifetime.

Japanese and German aggressions were, fortunately, even more short-lived. Japanese troops began occupying

Manchuria in September 1931; starting in 1937 the army took over several provinces in eastern China; starting in 1940 it took over Vietnam, Cambodia, Thailand, Burma, and all but a small part of what is today's Indonesia; and in June 1942 it occupied Attu (the westernmost island of Alaska's Aleutian chain) and Kiska island, about 300 kilometers to the east. These two easternmost outposts were lost just 13 months later, and the capitulation of Japan was signed on September 2, 1945; the imperial expansion thus lasted almost exactly 14 years. Meanwhile, Germany's Third Reich, which was meant to last for a thousand years, was gone 12 years and 3 months after Adolf Hitler became *Reichskanzler* on January 30, 1933.

And the American "empire"? Even if we were to believe in its actual existence and time its beginning in 1898 (the Spanish–American war and the takeover of the Philippines, Puerto Rico, and Guam), should we believe that it is still going strong? The Second World War was the last major conflict with a decisive US victory; the rest (Korean War, Vietnam War, Afghanistan War, Iraq War) were hard-to-classify mixtures of (costly) defeats and mutual exhaustion. Even the brief Gulf War of 1990–91 was no obvious win, as it led directly (12 years later) to the invasion and the stalemate of bloody years (2003–2011) in Iraq. And the country's share of the global economic product has been declining steadily since its unnatural peak in 1945 (when all other major economies were either destroyed or exhausted by war), and too many countries in the supposed American imperial orbit have shown little inclination to consent and follow. Clearly not an "empire" whose duration could be timed.

And who should pay the closest attention to these lessons of imperial demise? Obviously the Chinese Communist Party that is trying to suppress Tibet and Xinjiang, whose policies have not earned it any genuine friends along the country's long borders and have led to overreaching into the South China Sea, and whose decision to invest heavily (Silk Road–style) in poorer Asian and African countries is to buy long-term political influence. The party celebrated 70 years of the latest reincarnation of imperial rule in October 2019: given the history of modern imperial longevity, what are the odds that it will be around 70 years from now?

MACHINES, DESIGNS, DEVICES

Inventions That Made Our Modern World

How the 1880s created our modern world

According to the worshippers of the e-world, the late 20th century and the two opening decades of the 21st century brought us an unprecedented number of profound inventions. But that is a categorical misunderstanding, as most recent advances have been variations on two older fundamental discoveries: microprocessors (see INVENTING INTEGRATED CIRCUITS, p. 121) and exploiting radio waves, part of the electromagnetic spectrum. More powerful and more specialized microchips are now running everything from industrial robots and the autopilots of jetliners to kitchen ranges and digital cameras, and the most popular global brand for mobile communications has been using ultra high-frequency radio waves.

In fact, perhaps the most inventive time in human history was the 1880s. Have any two sets of primary inventions and epochal discoveries shaped the modern world more than electricity and internal combustion engines?

Electricity alone, without microchips, is enough to make a sophisticated and affluent world (we had one in the 1960s). Yet a microchip-governed e-world is utterly dependent on an electricity supply whose fundamental design remains beholden to thermal- and hydropower-generation systems, both of which reached the commercial market

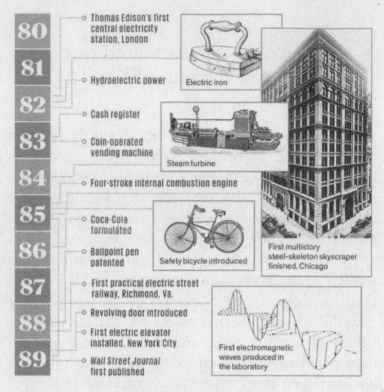

80	Thomas Edison's first central electricity station, London
81	Hydroelectric power
82	Cash register
83	Coin-operated vending machine
84	Four-stroke internal combustion engine
85	Coca-Cola formulated
86	Ballpoint pen patented
87	First practical electric street railway, Richmond, Va.
88	Revolving door introduced / First electric elevator installed, New York City
89	*Wall Street Journal* first published

Electric iron

Steam turbine

Safety bicycle introduced

First multistory steel-skeleton skyscraper finished, Chicago

First electromagnetic waves produced in the laboratory

The miraculous 1880s

in 1882 and still provide more than 80 percent of the world's electricity. And we aspire to make it available at least 99.9999 percent of the time, so that it can serve as the cornerstone of everything electronic.

Add to that the feats of Benz, Maybach, and Daimler, whose success with gasoline-fueled engines inspired Rudolf Diesel to come up with a more efficient alternative just a decade later (see WHY YOU SHOULDN'T WRITE

DIESEL OFF JUST YET, p. 109). By the end of the 19th century we also had conceptual designs of the most efficient of all internal combustion engines: the gas turbine. And it was in the 1880s when Heinrich Hertz's experiments proved the existence of electromagnetic waves (produced by oscillation of electric and magnetic fields), whose wavelengths increase from very short cosmic rays to X-rays, ultraviolet, visible, and infrared radiation, and to microwaves and radio waves. Their existence had been predicted by James Clerk Maxwell decades earlier, but Hertz made the practical opening onto our wireless world.

But the 1880s are also embedded in our lives in many smaller ways. Over a decade ago, in *Creating the Twentieth Century*, I traced several daily American experiences through mundane artifacts and actions that stem from that miraculous decade. A woman wakes up today in an American city and makes a cup of Maxwell House coffee (launched in 1886). She considers eating her favorite Aunt Jemima pancakes (sold since 1889) but goes for packaged Quaker Oats (available since 1884). She touches up her blouse with an electric iron (patented in 1882), applies antiperspirant (available since 1888), but cannot pack her lunch because she has run out of brown paper bags (the process to make strong kraft paper was commercialized in the 1880s).

She commutes on the light rail system (descended directly from the electric streetcars that began serving US cities in the 1880s), is nearly run over by a bicycle (the modern version of which—with equal-sized wheels and a chain drive—was another creation of the 1880s: see ENGINES ARE OLDER THAN BICYCLES!, p. 185), then goes through a revolving door (introduced in a Philadelphia building in 1888) into a multistory steel-skeleton skyscraper (the first one was finished in Chicago in 1885). She stops at a newsstand on the first floor, buys a copy of the *Wall Street Journal* (published since 1889) from a man who rings it up on his cash register (patented in 1883). Then she goes up to the 10th floor in an elevator (the first electric one was installed in a New York City building in 1889), stops at a vending machine (introduced, in its modern form, in 1883), and buys a can of Coca-Cola (formulated in 1886). Before she starts her work she jots down some reminders with her ballpoint pen (patented in 1888).

The 1880s were miraculous; they gave us such disparate contributions as antiperspirants, inexpensive lights, reliable elevators, and the theory of electromagnetism—although most people lost in their ephemeral tweets and in Facebook gossip are not even remotely aware of the true scope of this quotidian debt.

How electric motors
power modern civilization

Electrical devices advanced by leaps and bounds in the 1880s—the decade of the first power plants, durable lightbulbs, and transformers—but for most of the time advances in electric motors lagged behind.

Rudimentary direct current (DC) motors date back to the 1830s, when Thomas Davenport of Vermont patented the first American motor and used it to run a printing press, and Moritz von Jacobi of St. Petersburg used his motors to power a small paddle-wheel boat on the Neva River. But those battery-powered devices couldn't compete with steam power. More than a quarter-century passed before Thomas Edison finally commercialized a stencil-making electric pen to duplicate office documents; it, too, was powered by a DC motor. As commercial electricity generation began to spread after 1882, electric motors became common, and by 1887 US manufacturers were selling about 10,000 units a year, some of them operating the first electric elevators. All of them, however, ran on DC.

It fell to Edison's former employee, the Serbian-born Nikola Tesla, to set up a company of his own to develop a motor that could run on alternating current (AC). The goals were economy, durability, ease of operation, and

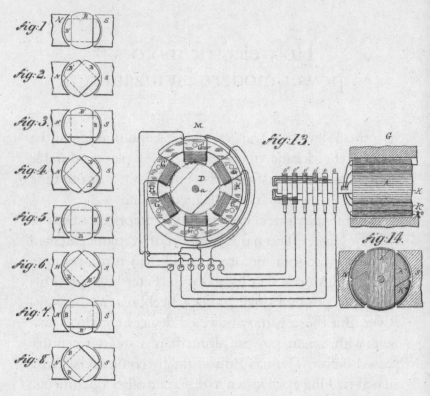

Illustrations appended to Tesla's US patent for an AC electric motor

safety. But Tesla was not the first to go public: in March 1888, the Italian engineer Galileo Ferraris gave a lecture on AC motors to the Royal Academy of Science in Turin, and published his findings a month later. This was a month before Tesla's corresponding lecture at the American Institute of Electrical Engineers. However, it was Tesla, helped with generous financing from US investors, who designed not only the AC induction motors but also the requisite AC transformers and distribution system. The two basic patents for his polyphase motor were granted in 1888. He filed some three dozen more by 1891.

In a polyphase motor, each electromagnetic pole in the stator (the stationary housing) has multiple windings, each of which carries alternating currents of equal frequency and amplitude but differing in phase from each other (in a three-phase motor by a third of a period).

George Westinghouse acquired Tesla's AC patents in July 1888. A year later, Westinghouse Co. began selling the world's first small electrical appliance: a fan powered by a 125-watt AC motor. Tesla's first patent was for a two-phase motor; modern households now rely on many

Nikola Tesla as a young man

small, single-phase AC electric motors; and the larger, more efficient three-phase machines are common in industrial applications. Mikhail Osipovich Dolivo-Dobrovolsky, a Russian engineer working as the chief electrician for Germany's AEG, built the first three-phase induction motor in 1889.

Today, some 12 billion small, nonindustrial motors are sold every year, including about 2 billion tiny (as small as 4 millimeters in diameter) DC devices used for cellphone vibration alerts, whose power requirements come to only a small fraction of a watt. At the other end of the spectrum are 6.5- to 12.2-megawatt motors powering French rapid trains (TGV), while the largest stationary motors in use for power compressors, fans, and conveyors have capacities exceeding 60 megawatts. This combination of ubiquity and power range makes it clear that electric motors are truly indispensable energizers of modern civilization.

Transformers—the unsung silent, passive devices

I have always disliked exaggerated claims of imminent scientific and technical breakthroughs, such as inexpensive fusion, cheap supersonic travel, and the terraforming of other planets. But I am fond of the simple devices that do so much of the fundamental work of modern civilization, particularly those that do so modestly—or even invisibly.

No device fits this description better than a transformer. Non-engineers may be vaguely aware that such

The world's largest transformer: Siemens for China

devices exist, but they have no idea how they work and how utterly indispensable they are for everyday life.

The theoretical foundation was laid in the early 1830s, with the independent discovery of electromagnetic induction by Michael Faraday and Joseph Henry. They showed that a changing magnetic field can induce a current of a higher voltage (known as "stepping up") or a lower one ("stepping down"). But it took another half-century before Lucien Gaulard, John Dixon Gibbs, Charles Brush, and Sebastian Ziani de Ferranti could design the first useful transformer prototypes. Next, a trio of Hungarian engineers—Ottó Bláthy, Miksa Déri, and Károly Zipernowsky—improved the design by building a toroidal (donut-shaped) transformer, which they exhibited in 1885.

The very next year, a better design was introduced by a trio of American engineers—William Stanley, Albert Schmid, and Oliver B. Shallenberger, who were working for George Westinghouse. The device soon assumed the form of the classic Stanley transformer that has been retained ever since: a central iron core made of thin silicon steel laminations, one part shaped like an "E" and the other shaped like an "I" to make it easy to slide pre-wound copper coils into place.

In his address to the American Institute of Electrical Engineers in 1912, Stanley rightly marveled at how the device provided "such a complete and simple solution for a difficult problem. It so puts to shame all mechanical attempts at regulation. It handles with such ease, certainty, and economy vast loads of energy that are instantly given to or taken from it. It is so reliable, strong, and certain. In this mingled steel and copper, extraordinary forces are so nicely balanced as to be almost unsuspected."

The biggest modern incarnations of this enduring design have made it possible to deliver electricity across great distances. In 2018, Siemens delivered the first of seven record-breaking 1,100-kilovolt transformers that will enable electricity supply to several Chinese provinces linked to a nearly 3,300-kilometer-long, high-voltage DC line.

The sheer number of transformers has risen above anything Stanley could have imagined, thanks to the explosion of portable electronic devices that have to be charged. In 2016, the global output of smartphones alone was in excess of 1.8 billion units, each one supported by a charger housing a tiny transformer. You don't have to take your phone charger apart to see the heart of that small device: a complete iPhone charger teardown is posted on the Internet, with the transformer as one of its largest components.

But many chargers contain even tinier transformers. These are non-Stanley (that is, not wire-wound) devices that take advantage of the piezoelectric effect—the ability of a strained crystal to produce a current, and of a current to strain or deform a crystal. Sound waves impinging on such a crystal can produce a current, and a current flowing through such a crystal can produce sound. One current can in this way be used to create another current of a very different voltage.

And the latest innovation is electronic transformers. They are much reduced in volume and mass compared with traditional units, and they will become particularly important for integrating intermittent sources of electricity—wind and solar—into the grid and for enabling DC microgrids.

Why you shouldn't write diesel off just yet

On February 17, 1897, Moritz Schröter, a professor of theoretical engineering at Technische Universität in Munich, conducted the official certification test of Rudolf Diesel's new engine. The goal of the test was to verify the machine's efficiency and hence to demonstrate its suitability for commercial development.

The 4.5-metric-ton engine performed impressively: at its full power of 13.4 kilowatts (18 horsepower, equal to a modern small motorcycle), its net efficiency reached 26 percent, much better than any contemporary gasoline engine. With obvious pride, Diesel wrote to his wife, "Nobody's engine design has achieved what mine has done, and so I can have the proud awareness of being the first one in my specialty." Later in that year the engine's net efficiency reached 30 percent, making the machine twice as efficient as the gasoline-fueled Otto engines of the day.

Over time, that efficiency gap has narrowed, but today's diesel engines remain at least 15 to 20 percent more efficient than their gasoline-fueled rivals. Diesels have several advantages: they use fuel of a higher energy density (it contains nearly 12 percent more energy than the same volume of gasoline, and hence a vehicle can

No. 608,845.

Patented Aug. 9, 1898.

R. DIESEL.
INTERNAL COMBUSTION ENGINE.
(Application filed July 15, 1895.)

(No Model.)

2 Sheets—Sheet 1.

WITNESSES:
Jao. W. Thomas
Eugenie A. Aersider.

INVENTOR:
Rudolf Diesel,
BY
Alber du Krewp
ATTORNEY

Rudolf Diesel's US patent for his new internal combustion engine go further with the same tank volume); their self-ignition involves compression ratios twice as high as those in gasoline engines (resulting in a more complete combustion and in cooler exhaust gas); they can burn lower-quality, and hence cheaper, fuel; and modern electronic injection systems can spray the fuel into their cylinders at high pressures resulting in higher efficiencies and in cleaner exhaust.

But, disappointingly, in 1897 the record-setting test was not followed by rapid commercial deployment. Diesel's conclusion that he had "a thoroughly marketable machine" and that "the rest will develop automatically on its own worth" was wrong. Not until 1911 did the Danish vessel *Selandia* become the first oceangoing freighter powered by a diesel engine, and diesels dominated shipping only after the First World War. Heavy railroad traction was their first land conquest, followed by heavy road transport, off-road vehicles, and construction and agricultural machinery. The first diesel car, the Mercedes-Benz 260 D, came in 1936. Today, in the European Union, about 40 percent of all passenger cars are diesels but in the US (which has cheaper gasoline) diesel accounts for just 3 percent.

Rudolf Diesel's initial hope was to see small engines used primarily by small, independent producers as tools of industrial decentralization, but more than 120 years later, the very opposite is true. Diesels are the uncontested enablers of massively centralized industrial production and the irreplaceable prime movers of globalization.

Diesels power virtually all container ships and all carriers of vehicles and bulk commodities such as oil, liquefied natural gas, ores, cement, fertilizers, and grain. They also power nearly all trucks and freight trains.

Most of the items that readers of this book eat or wear are transported at least once, and usually many times, by diesel-powered machines, often from other continents: clothes from Bangladesh, oranges from South Africa, crude oil from the Middle East, bauxite from Jamaica, cars from Japan, computers from China. Without the low operating costs, high efficiency, high reliability, and great durability of diesel engines, it would have been impossible to reach the extent of globalization that now defines the modern economy.

Over more than a century of use, diesel engines have increased both in capacity and efficiency. The largest machines in shipping are now rated at more than 81 megawatts (109,000 horsepower), and their top net efficiency is just above 50 percent—better than that of gas turbines, which are at about 40 percent (see WHY GAS TURBINES ARE THE BEST CHOICE, p. 139).

And diesel engines are here to stay. There are no readily available mass-mover alternatives that could keep integrating the global economy as affordably, efficiently, and reliably as Diesel's machines.

Capturing motion—from horses to electrons

Eadweard Muybridge (1830–1904), an English photographer, established his American fame in 1867 by taking a mobile studio to Yosemite Valley and producing large silver prints of its stunning vistas. Five years later he was hired by Leland Stanford, then the president of the Central Pacific Railroad, formerly the governor of California, and latterly the founder of the eponymous

Muybridge's galloping horse

university in Palo Alto. Stanford—who was also a horse breeder—challenged Muybridge to settle the old dispute: whether all four of a horse's legs are off the ground when running.

Muybridge found it difficult to prove the point. In 1872 he took (and then lost) a single image of a trotting horse with all hooves aloft. But he persevered, and his eventual solution was to capture moving objects with cameras capable of a shutter speed as brief as 1/1,000 of a second.

The conclusive experiment took place on June 19, 1878, at Stanford's Palo Alto farm. Muybridge lined up thread-triggered glass-plate cameras along the track, used a white-sheet background for the best contrast, and copied the resulting images as a sequence of still photographs (silhouettes) on the disc of a simple circular device he called a zoopraxiscope, in which a rapid series of rotating stills conveyed motion.

Sallie Gardner, the horse Stanford had provided for the test, clearly had all four hooves off the ground at the gallop. But the airborne moment did not take place as portrayed in famous paintings, perhaps most notably Théodore Géricault's *The 1821 Derby at Epsom*, now hanging in the Louvre, which shows the animal's legs extended away from its body. Instead, it occurred when the horse's legs were beneath its body, just prior to the moment the horse pushed off with its hind legs.

This work led to Muybridge's *magnum opus*, which he prepared for the University of Pennsylvania. Starting in

1883, he began an extensive series depicting animal and human locomotion. Its creation relied on 24 cameras fixed in parallel to a 36-meter-long track with two portable sets of 12 batteries at each end. The track had a marked background, and animals or people activated the shutters by breaking stretched strings.

The final product was a book with 781 plates, published in 1887. This compendium showed not only running domestic animals (dogs and cats, cows and pigs) but also a bison, a deer, an elephant, and a tiger, as well as a running ostrich and a flying parrot. Human sequences depicted running, and also ascents, descents, lifts, throws, wrestling, a child crawling, and a woman pouring a bucket of water over another woman.

Muybridge's 1,000 frames a second soon became 10,000. By 1940, the patented design of a rotating mirror camera raised the rate to 1 million per second. In 1999, Ahmed Zewail won the Nobel Prize in Chemistry for developing a spectrograph that could capture the transition states of chemical reactions on a scale of femtoseconds—that is, 10^{-15} seconds, or one-millionth of one-billionth of a second.

Today, we can use intense, ultrafast laser pulses to capture events separated by mere attoseconds, or 10^{-18} seconds. This time resolution makes it possible to see what has until recently been hidden from any direct experimental access: the motions of electrons on the atomic scale.

Many examples can be given to illustrate the extraordinary scientific and engineering progress we have made since the closing decades of the 19th century, and several impressive cases—including the luminous efficacy of light (see WHY SUNLIGHT IS STILL BEST, p. 158) and the income- and performance-adjusted cost of electricity (see THE REAL COST OF ELECTRICITY, p. 170)—are detailed in this book, but the contrast between the discoveries of Eadweard Muybridge and Ahmed Zewail is as impressive as any other advance I can think of: from settling the dispute about airborne horse hooves to observing flitting electrons.

From the phonograph to streaming

When Thomas Edison died in 1931, at 84, he held almost 1,100 patents in the United States and more than 2,300 patents worldwide. By far the most famous one was his patent for the lightbulb, but he came up neither with the idea of an evacuated glass container nor with the use of an incandescing filament. More fundamental was Edison's conception, entirely *de novo*, of the complete system of electricity generation, transmission, and conversion,

Thomas Edison with his phonograph

which he put into operation first in London and then in lower Manhattan, in 1882.

But for sheer originality bordering on the magical, nothing compares to Edison's US Patent No. 200,521, issued on February 19, 1878, for the first-ever way to hear recorded sound.

The phonograph (a device for the mechanical recording and reproduction of sound) was born out of the telegraph and telephone. Edison spent years trying to improve the former—most of his early patents were related to printing telegraphs—and he had been intrigued by the latter ever since its introduction in 1876. Edison got his first telephone-related patents in 1878. He noticed that playing a recorded telegraph tape at a high speed produced noises resembling spoken words. What would happen if he recorded a telephone message by attaching a needle to the receiver's diaphragm, produced a pricked tape, and then replayed that tape? He designed a small device with a grooved cylinder overlaid with tinfoil that could easily receive and record the motions of the diaphragm. "I then shouted, 'Mary had a little lamb,' etc.," Edison later recalled. "I adjusted the reproducer, and the machine reproduced it perfectly. I was never so taken aback in my life. Everybody was astonished. I was always afraid of things that worked the first time."

Soon he took the phonograph on a tour, and even to the White House. His advertisement called it (incongruously) "Thomas Edison's Final Achievement," and the inventor's wish was that every American family might eventually buy the machine. He greatly improved its design during the late 1880s by using wax-coated cylinders (originally conceived by colleagues of Alexander Graham Bell, inventor of the telephone) and a battery-powered electric motor, marketing it as a recorder of family voices and a music box, as well as a dictation machine for businesses and an audio book for the blind.

However, sales were never spectacular. Wax cylinders, especially the early versions, were fragile, hard to make, and therefore expensive. By 1887, the American Graphophone Company had obtained the patent to a rival version of the device, yet it remained costly (equal to about $4,000 today).

During the 1880s, Edison was preoccupied with introducing and improving electric lights, and inventing and designing generation and transmission systems. But in 1898 he began selling the Edison Standard Phonograph for $20, or about $540 in today's monies. A year later came an inexpensive Gem model for just $7.50 (Sears, Roebuck & Co. was selling an iron bed for about that much). But by the time Edison was mass-producing unbreakable celluloid cylinders in 1912, shellac disc records for the gramophone (first patented by Emile Berliner in 1887) had taken over.

Edison always found it hard to let go of his early inventions. The last phonograph cylinders were made in October 1929. Flat discs with a spiral groove, used on the gramophone, remained dominant for most of the

20th century, until new modes of sound recording came in quick succession. US sales of LPs peaked in 1978, compact cassettes did so a decade later, and then CDs—introduced in 1984—peaked in 1999. Those sales were cut in half just seven years later, and they are now surpassed by music downloads, including free wireless streaming. How would Edison have regarded these dematerialized methods for reproducing sound?

Inventing integrated circuits

In 1958, 11 years after Bell Labs reinvented the transistor, it became clear that semiconductors would be able to conquer the electronics market only if they could be greatly miniaturized. There wasn't much progress to be made by hand-soldering separate components into circuits, but as is often the case, the solution came just when it was needed.

In July 1958, Jack S. Kilby of Texas Instruments came up with the monolithic idea. His patent application described it as "a novel miniaturized electronic circuit fabricated from a body of semiconductor material containing a diffused p-n junction wherein all components of the electronic circuit are completely integrated into the body of semiconductor material." And Kilby stressed that "there is no limit upon the complexity or configuration of circuits which can be made in this manner."

The idea was perfect, but its execution—as depicted in Kilby's February 1959 patent application—was unworkable, because it had the wire connections raising, arch-like, above the wafer's surface, hardly the way to make a planar component. Kilby knew that this wouldn't work, and that is why he added a note about connections to be made in other ways. As an example, he mentioned gold

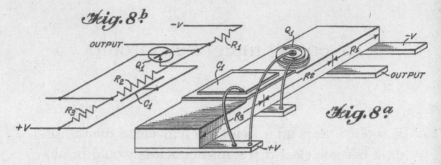

Integrated circuit: Kilby's "flying wires" patent

deposited on the thin silicon oxide layer on the wafer's surface.

Unbeknownst to him, in January 1959, Robert Noyce, then the director of research at Fairchild Semiconductor Corporation, was to jot down in his lab notebook an improved version of the very same idea. "It would be desirable to make multiple devices on a single piece of silicon, in order to be able to make interconnections between devices as part of the manufacturing process, and thus reduce size, weight, etc. as well as cost per active element," Noyce wrote. Moreover, the drawing accompanying Noyce's July 1959 patent application contained no flying wires; instead, it clearly depicted a planar transistor and "leads in the form of vacuum-deposited or otherwise formed metal strips extending over and adherent to the insulating oxide layer for making electrical connections to and between various regions of the semiconductor body without shorting the junctions."

Noyce's patent was granted in April 1961; Kilby's in

July 1964. The litigation went all the way to the Supreme Court, which in 1970 refused to hear the case, upholding a lower court's ruling of Noyce's priority. That decision made no practical difference, because in 1966 the two companies had agreed to share their production licenses, and the origins of the integrated circuit became yet another outstanding example of concurrent independent inventions. The basic conceptual idea was identical; both inventors received the National Medal of Science, and both were inducted into the National Inventors Hall of Fame. Noyce lived only to 62, but Kilby survived to share a Nobel Prize in Physics in 2000 at the age of 77, five years before his death.

Texas Instruments called the new designs "Micrologic

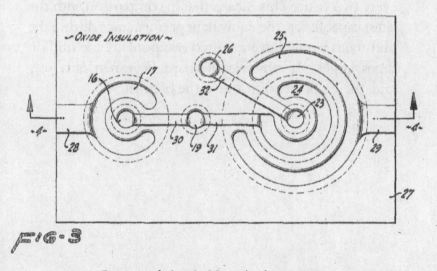

Integrated circuit: Noyce's planar patent

elements." They were chosen to control intercontinental ballistic missiles and to help land men on the Moon.

Their subsequent progress, captured by the still-enduring Moore's Law (see MOORE'S CURSE: WHY TECHNICAL PROGRESS TAKES LONGER THAN YOU THINK, p. 125), has been one of the defining developments of our time. By 1971, basic integrated circuits had matured into simple microprocessors with thousands of components, which then advanced to designs that made personal computing affordable, starting in the mid-1980s. By 2003, the component total had surpassed 100 million, and by 2015 it reached 10 billion transistors. That represents an aggregate growth of eight orders of magnitude since 1965, averaging about 37 percent a year, with the number of components on a given area doubling about every two years. This means that in comparison with the latest capabilities, the equivalent performance during the mid-1960s would have required components 100 million times larger. As physicist Richard Feynman famously said, there's plenty of room at the bottom.

Moore's Curse: Why technical progress takes longer than you think

In 1965, Gordon Moore—at that time the director of R&D at Fairchild Semiconductor—noted that "the complexity for minimum component costs has increased at a rate of roughly a factor of two per year . . . Certainly over the short term this rate can be expected to continue, if not to increase." Over the longer term, the doubling rate settled at about two years, or an exponential growth rate of 35 percent a year. This is Moore's Law.

As components have gotten smaller, denser, faster, and cheaper, they have increased the power and cut the costs of many products and services, notably computers and mobile phones. The result has been a revolution in electronics.

But this revolution has been both a blessing and a curse, for it has had the unintended effect of raising expectations for technical progress. We are assured that rapid progress will soon bring self-driving electric cars, individually tailored cancer cures, and instant 3D printing of hearts and kidneys. We are even told it will pave the way for the world's transition from fossil fuels to renewable energies.

But the doubling time for transistor density is no guide to technical progress generally. Modern life depends

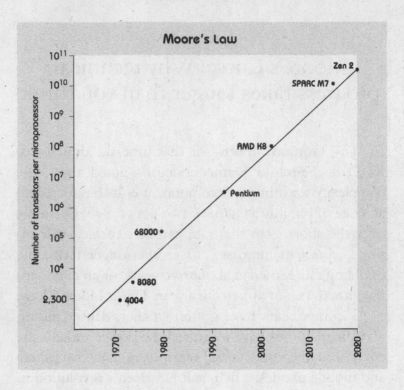

Moore's Law

on many processes that improve rather slowly, not least the production of food and energy and the transportation of people and goods—and such slow rates prevail not only for pre-1950 advances but also for essential improvements and innovation coinciding with the development of transistors (their first commercial application was in hearing aids in 1952).

Corn, America's leading crop, has seen its average yields rising by 2 percent a year since 1950. Yields of rice, China's largest staple, have been going up by about

1.6 percent during the past 50 years. The efficiency with which steam turbogenerators convert thermal power to electricity generation rose annually by about 1.5 percent during the 20th century; if you instead compare the steam turbogenerators of 1900 with the combined-cycle power plants of 2000 (which mate gas turbines to steam boilers), that annual rate increases to 1.8 percent. Advances in lighting have been more impressive than in any other sector of electricity conversion, but between 1881 and 2014 light efficacy (lumens per watt) rose by just 2.6 percent a year for indoor lights, and by 3.1 percent for outdoor lighting (see WHY SUNLIGHT IS STILL BEST, p. 158).

The speed of intercontinental travel rose from about 35 kilometers per hour for large ocean liners in 1900 to 885 km/h for the Boeing 707 in 1958, an average rise of 5.6 percent a year. But speed of jetliners has remained essentially constant ever since—the Boeing 787 cruises just a few percent faster than the 707. Between 1973 and 2014, the fuel-conversion efficiency of new US passenger cars (even after excluding monstrous SUVs and pickups) rose at an annual rate of just 2.5 percent, from 13.5 to 37 miles per gallon (that's from 17.4 to 6.4 liters per 100 kilometers). And finally, the energy cost of steel (coke, natural gas, and electricity), our civilization's most essential metal, was reduced from about 50 gigajoules per ton to less than 20 between 1950 and 2010—that is, an annual rate of about −1.7 percent.

Energy, material, and transportation fundamentals

that enable the functioning of modern civilization and that circumscribe its scope of action are improving steadily but slowly. Gains in performance mostly range from 1.5 to 3 percent a year, as do the declines in cost.

And so, outside the microchip-dominated world, innovation simply does not obey Moore's Law, proceeding at rates that are lower by an order of magnitude.

The rise of data: Too much too fast

Once upon a time, information was deposited only inside human brains, and ancient bards could spend hours retelling stories of conflicts and conquests. Then external data storage was invented.

Small clay cylinders and tablets, invented in Sumer in southern Mesopotamia some 5,000 years ago, often contained just a dozen cuneiform characters in that ancient language, equivalent to a few hundred (or 10^2) bytes. The *Oresteia*, a trilogy of Greek tragedies written by Aeschylus in the fifth century BCE, amounts to about 300,000 (or 10^5) bytes. Some rich senators in imperial Rome had libraries housing hundreds of scrolls, with one large collection holding at least 100 megabytes (10^8 bytes).

A radical shift came with Johannes Gutenberg's printing press, using movable type. By 1500, less than half a century after its introduction, European printers had released more than 11,000 new book editions. This extraordinary rise was joined by advances in other forms of stored information. First came engraved and woodcut music scores, illustrations, and maps. Then, in the 19th century, photographs, sound recordings, and movies. New information storage modes added during the 20th

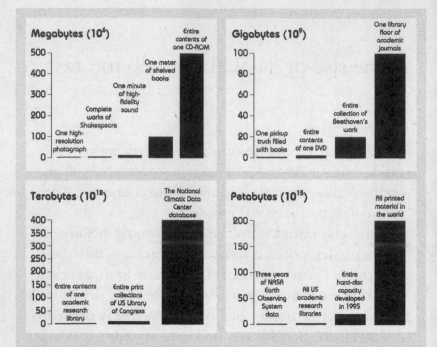

Megabytes (10⁶)

- One high-resolution photograph
- Complete works of Shakespeare
- One minute of high-fidelity sound
- One meter of shelved books
- Entire contents of one CD-ROM

(scale: 0, 100, 200, 300, 400, 500)

Gigabytes (10⁹)

- One pickup truck filled with books
- Entire contents of one DVD
- Entire collection of Beethoven's work
- One library floor of academic journals

(scale: 0, 20, 40, 60, 80, 100)

Terabytes (10¹²)

- Entire contents of one academic research library
- Entire print collections of US Library of Congress
- The National Climatic Data Center database

(scale: 0, 50, 100, 150, 200, 250, 300, 350, 400)

Petabytes (10¹⁵)

- Three years of NASA Earth Observing System data
- All US academic research libraries
- Entire hard-disc capacity developed in 1995
- All printed material in the world

(scale: 0, 50, 100, 150, 200)

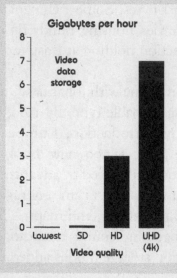

Gigabytes per hour

Video data storage

(scale: 0, 1, 2, 3, 4, 5, 6, 7, 8)

- Lowest
- SD
- HD
- UHD (4k)

Video quality

century included magnetic tapes and long-playing records and, beginning in the 1960s, computers expanded the scope of digitization to medical imaging (a digital mammogram is 50 megabytes), animated movies (2–3 gigabytes), intercontinental financial transfers, and eventually the mass emailing of spam (more than 100 million messages sent every minute). Such digitally stored information rapidly surpassed all printed materials. Shakespeare's plays and poems in their entirety amount to 5 megabytes, the equivalent of just a single high-resolution photograph, or of 30 seconds of high-fidelity sound, or of 8 seconds of streamed high-definition video.

Printed materials have thus been reduced to a marginal component of overall global information storage. By the year 2000, all books in the Library of Congress held more than 10^{13} bytes (more than 10 terabytes), but that was less than 1 percent of the total collection (10^{15} bytes, or about 3 petabytes) once all photographs, maps, movies, and audio recordings were added.

And in the 21st century, information is being generated ever faster. In its latest survey of data generated per minute in 2018, Domo, a cloud service, listed more than 97,000 hours of video streamed by Netflix users, nearly 4.5 million videos watched on YouTube, just over 18 million forecast requests on the Weather Channel, and more than 3 quadrillion bytes (3.1 petabytes) of other Internet data used in the United States alone. By 2016, the annual global data-creation rate surpassed 16 zettabytes (1 ZB is 10^{21} bytes), and by 2025 it is expected to rise by another order of magnitude—that is, to about 160 zettabytes (10^{23} bytes). And according to Domo, in 2020, 1.7 megabytes of data is generated every second for every one of the world's nearly 8 billion people.

These quantities lead to some obvious questions. Only a fraction of the data flood can be stored, but which part should that be? Challenges of storage are obvious even if less than 1 percent of this flow gets preserved. And for whatever we decide to store, the next question is for how long the data should be preserved. No storage need last forever, but what is the optimal span?

The highest prefix in the international system of units (in which thousand is k = 10^3, and million is M = 10^6) is yotta- (Y = 10^{24}, or a trillion trillions). We'll have that many bytes within a decade, and it will be increasingly difficult to evaluate it—even if those tasks are increasingly left to machines. And once we start creating more than 50 trillion bytes of information per person per year, will there be any real chance of making effective use of it? After all, there are fundamental differences between accumulated data, useful information, and insightful knowledge.

Being realistic about innovation

Modern societies are obsessed with innovation. At the end of 2019, Google searches returned 3.21 billion hits for "innovation," easily beating "terrorism" (481 million), "economic growth" (about 1 billion), and "global warming" (385 million). We are to believe that innovation will open every conceivable door: to life expectancies far beyond 100 years, to the merging of human and machine consciousness, to essentially free solar energy.

This uncritical genuflection before the altar of innovation is wrong on two counts: It ignores those big, fundamental quests that have failed after spending huge sums on research. And it has little to say about why we so often stick to an inferior practice even when we know there's a superior course of action.

The fast breeder reactor, so called because it produces more nuclear fuel than it consumes, is one of the most remarkable examples of a prolonged and costly innovation failure. In 1974, General Electric predicted that by 2000 about 90 percent of the United States' electricity would come from fast breeders. GE was merely reflecting a widespread expectation: during the 1970s, the governments of France, Japan, the Soviet Union, the United Kingdom, and the United States were all

Prototype of a maglev train unveiled by the China Railway
Rolling Stock Corporation in 2019

investing heavily in the development of breeders. But
high costs, technical problems, and environmental concerns led to shutdowns of British, French, Japanese,
American (and also smaller German and Italian) programs, while China, India, Japan, and Russia are still
operating experimental reactors. After the world as a
whole spent well above $100 billion in today's money
over some six decades of effort, there has been no real
commercial payoff.

Other promised fundamental innovations that are
still not commercial concerns include hydrogen (fuel
cell) powered cars, magnetic levitation (maglev) trains,
and thermonuclear energy. The last one is perhaps the

most notorious example of an ever-receding innovative achievement.

The second category of failed innovations—things we keep on doing even though we know we shouldn't—range from quotidian practices to theoretical concepts.

Two annoying examples are daylight saving time and boarding airplanes. Why do we keep imposing "daylight saving time" changes semiannually (energy savings a justification of the switch) when we know they don't really save anything? And commercial flights now take longer to board than during the 1970s despite the fact that we know a number of methods that are faster than the current inefficient favorites. For example, we might be seating people in a reverse pyramid, alternately boarding them at the back and at the front simultaneously (spreading things out to minimize bottlenecks), or simply by abolishing assigned seating.

And why do we measure the progress of economies by gross domestic product? GDP is simply the total annual value of all goods and services transacted in a country. It rises not only when lives get better and economies progress but also when bad things happen to people or to the environment. Higher alcohol sales, more driving under the influence, more accidents, more emergency-room admissions, more injuries, more people in jail—GDP goes up. More illegal logging in the tropics, more deforestation and biodiversity loss, higher timber sales—again, GDP goes up. We know better, but we still worship high annual GDP growth rate, regardless of where it comes from.

Human minds have many irrational preferences: we love to speculate about wild and crazy innovations but cannot be bothered to fix common challenges by relying on practical innovation waiting to be implemented. Why do we not improve the boarding of planes rather than delude ourselves with visions of hyperloop trains and eternal life?

FUELS AND ELECTRICITY
Energizing Our Societies

Why gas turbines are the best choice

In 1939, the world's first industrial gas turbine began to generate electricity in a municipal power station in Neuchâtel, Switzerland. The machine, installed by Brown Boveri, vented the exhaust without making use of its heat, and the turbine's compressor consumed nearly three-quarters of the generated power. That resulted in an efficiency of just 17 percent, or about 4 megawatts.

The disruption of the Second World War and the economic difficulties that followed made the Neuchâtel turbine a pioneering exception until 1949, when Westinghouse and General Electric introduced their first designs of limited power. There was no rush to install them, as the market was dominated by large coal-fired plants that generated the least expensive electricity. By 1960, the most powerful gas turbine reached 20 megawatts, still an order of magnitude smaller than the output of most steam turbogenerators.

In November 1965, the great power blackout in the Northeastern United States changed many minds: gas turbines could operate at full load within minutes. But rising oil and gas prices and a slowing demand for electricity prevented any rapid expansion of the new technology. The shift came only during the late 1980s; by 1990,

Interior of a large gas turbine

almost half of all new installed US electricity generating capacity was in gas turbines of increasing power, reliability, and efficiency.

But even efficiencies in excess of 40 percent produce exhaust gases of about 600°C, hot enough to generate steam in an attached steam turbine. This tandem of a gas turbine and a steam turbine—combined cycle gas turbine (CCGT)—was first developed in the late 1960s, and the best efficiencies of CCGTs now top 60 percent. No other prime mover is less wasteful.

Siemens currently offers a CCGT for utility generation rated at 593 megawatts, nearly 40 times as powerful as the Neuchâtel machine and operating at 63 percent efficiency. General Electric's 9HA gas turbine delivers 571 megawatts when operating alone (simple-cycle generation) and 661 megawatts (63.5 percent efficiency) when coupled with a steam turbine (CCGT).

Gas turbines are the ideal suppliers of peak power and the best backups for intermittent wind and solar generation. In the United States, they are now by far the most affordable choice for new generating capacities. The levelized cost of electricity (a measure of the lifetime cost of an energy project) for new capacities entering service in 2023 is forecast to be about $60 per megawatt-hour for coal-fired steam turbogenerators with partial carbon capture, $48/MWh for solar photovoltaics, and $40/MWh for onshore wind—but less than $30/MWh for conventional gas turbines and less than $10/MWh for CCGTs.

Gas turbines are also used around the world for the combined production of electricity and heat. Steam and hot water are required in many industries, and are used to energize central heating systems that are particularly common in many large European cities. These turbines have even been used to heat and light extensive Dutch greenhouses, which additionally benefit from the carbon dioxide that is generated—as it speeds up the growth of vegetables. Gas turbines also run compressors in many industrial enterprises and in the pumping stations of long-distance pipelines.

The verdict is clear: no other combustion machines combine so many advantages as do modern gas turbines. They're compact, easy to transport and install, and relatively silent, affordable, and efficient, offering nearly instant output and able to operate without water cooling. All this makes them the unrivaled machine to supply both mechanical energy and heat.

And their longevity? The Neuchâtel turbine was decommissioned in 2002, after 63 years of operation—not due to any failure in the machine but because of a damaged generator.

Nuclear electricity—an unfulfilled promise

The age of commercial nuclear electricity generation began on October 17, 1956, when Queen Elizabeth II switched on Calder Hall, on the northwest coast of England. Sixty years is long enough to judge the technology, and I still cannot improve on my evaluation from over a decade ago: a "successful failure."

The success part is well-documented. After a slow start, reactor construction began to accelerate during the late 1960s, and by 1977 more than 10 percent of US electricity came from fission, rising to 20 percent by 1991. That was a faster penetration of the market than photovoltaics and wind turbines have managed since the 1990s.

In late 2019, the world had 449 operating reactors (and 53 under construction), many with capacity factors of better than 90 percent. That's the share of the reactors' potential output that they averaged year-round, producing more than twice as much electricity as photovoltaic cells and wind turbines combined. In 2018, nuclear power provided the highest share of electricity in France (about 72 percent), 50 percent in Hungary, Swiss reactors contributed 38 percent, and in South Korea it was 24 percent, while the share in the US was just below 20 percent.

Number of operable nuclear reactors

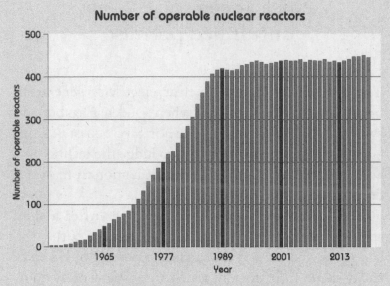

Nuclear electricity production

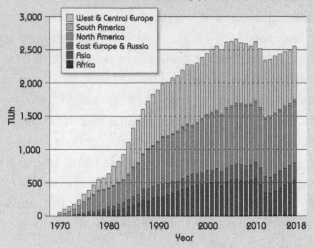

The "failure" part has to do with unmet expectations. The claim that nuclear electricity would be "too cheap to meter" is not apocryphal: that's what Lewis L. Strauss, chairman of the United States Atomic Energy Commission in 1954, told the National Association of Science Writers in New York in September of that year. And equally audacious claims were still to come. In 1971, Glenn Seaborg, a Nobelist and the then-chairman of the Atomic Energy Commission, predicted that nuclear reactors would generate nearly all the world's electricity by 2000. Seaborg envisioned giant coastal "nuplexes" desalinating sea water; geostationary satellites powered by compact nuclear reactors for broadcasting TV programs; nuclear-powered tankers; and nuclear explosives that would alter the flow of rivers and excavate underground cities. Meanwhile, nuclear propulsion would carry men to Mars.

But the project to generate electricity from fission stalled during the 1980s, as demand for electricity in affluent economies fell and problems with nuclear power plants multiplied. And three failures were worrisome: accidents at Three Mile Island in Pennsylvania in 1979, at Chernobyl in Ukraine in 1986, and at Fukushima in Japan in 2011 provided further evidence for those opposed to fission under any circumstances.

Meanwhile, there have been cost overruns in the construction of nuclear plants and a frustrating inability to come up with an acceptable way for permanent storage of the spent nuclear fuel (currently temporarily stored in containers at power plant sites). Nor has there been much success in switching to reactors that might be safer and less expensive than the dominant design of pressurized water reactors, which are essentially beached versions of US Navy submarine designs from the 1950s.

As a result, the Western public remains unconvinced, electricity-generating companies are wary, Germany and Sweden are on course to shutting down their entire industries, and even France plans to cut back. The reactors that are now under construction worldwide will not be able to make up for the capacity that will be lost as aging reactors are shut down in coming years.

The only leading economies with major expansion plans are in Asia, led by China and India, but even they can't do much to reverse the decline in the share of nuclear power in worldwide electricity generation. That share peaked at nearly 18 percent in 1996, fell to 10 percent in 2018, and is expected to bump up to just 12 percent by 2040, according to the International Energy Agency.

There are many things we could do—above all, use better reactor designs and act resolutely on waste storage—to generate a significant share of electricity from nuclear fission and so limit carbon emissions. But that would require an unbiased examination of the facts, and a truly long-range approach to global energy policy. I see no real signs of either.

Why you need fossil fuels to get electricity from wind

Wind turbines are the most visible symbols of the quest for renewable electricity generation. And yet, although they exploit the wind, which is as free and as green as energy can be, the machines themselves are pure embodiments of fossil fuels.

Large trucks bring steel and other raw materials to the site, earth-moving equipment beats a path to otherwise-inaccessible high ground, large cranes erect the structures—and all these machines burn diesel fuel. So do the freight trains and cargo ships that convey the materials needed for the production of cement, steel, and plastics. For a 5-megawatt turbine, the steel alone averages 150 tons for the reinforced concrete foundations, 250 tons for the rotor hubs and nacelles (which house the gearbox and generator), and 500 tons for the towers.

If wind-generated electricity were to supply 25 percent of global demand by 2030, then even with a high average capacity factor of 35 percent the aggregate installed wind power of about 2.5 terawatts would require roughly 450 million tons of steel. And that's without counting the metal for towers, wires, and transformers for the new high-voltage transmission links that would be needed to connect it all to the grid.

Large plastic blade of a modern wind turbine: difficult to make, more difficult to transport, even more difficult to recycle

A lot of energy goes into making steel. Sintered or pelletized iron ore is smelted in blast furnaces, charged with coke made from coal, and receives infusions of powdered coal and natural gas. Pig iron (iron made in blast furnaces) is decarbonized in basic oxygen furnaces. Then steel goes through continuous casting processes (which turn molten steel directly into the rough shape of the final product). Steel used in turbine construction typically embodies about 35 gigajoules per ton.

To make the steel required for wind turbines that might operate by 2030, you'd need fossil fuels equivalent to more than 600 million tons of coal.

A 5-megawatt turbine has three roughly 60-meter-long airfoils, each weighing about 15 tons. They have light balsa or foam cores and outer laminations made mostly from glass-fiber-reinforced epoxy or polyester resins. The glass is made by melting silicon dioxide and other mineral oxides in furnaces fired by natural gas. The resins begin with ethylene derived from light hydrocarbons—most commonly the products of naphtha cracking, liquefied petroleum gas, or the ethane in natural gas.

The final fiber-reinforced composite embodies on the order of 170 gigajoules per ton. Therefore, to get 2.5 terawatts of installed wind power by 2030, we would need an aggregate rotor mass of about 23 million tons, incorporating the equivalent of about 90 million tons of crude oil. And when all is in place, the entire structure must be waterproofed with resins whose synthesis starts with ethylene. Yet another required oil product is lubricant, for the turbine gearboxes, which has to be changed periodically during the machine's two-decade lifetime.

Undoubtedly, in less than a year a well-sited and well-built wind turbine will generate as much energy as it took to produce it. However, all of it will be in the form of intermittent electricity—while its production, installation, and maintenance remain critically dependent on specific fossil energies. Moreover, for most of these energies—coke for iron-ore smelting; coal and petroleum coke to fuel cement kilns; naphtha and natural gas as feedstock and fuel for the synthesis of plastics and the making of fiberglass; diesel fuel for ships, trucks, and construction machinery; lubricant for gearboxes—we have no non-fossil substitutes that would be readily available on the requisite large commercial scales.

For a long time to come—until all energies used to produce wind turbines and photovoltaic cells come from renewable energy sources—modern civilization will remain fundamentally dependent on fossil fuels.

How big can a wind turbine be?

Wind turbines have certainly grown up. When the Danish firm Vestas began the trend toward gigantism in 1981, its three-blade machines were capable of a mere 55 kilowatts. That figure rose to 500 kilowatts in 1995, reached 2 megawatts in 1999, and today stands at 5.6 megawatts. In 2021, MHI Vestas Offshore Wind's V164 will stand 105 meters high at the hub, swing 80-meter blades, and generate up to 10 megawatts, making it the first commercially available double-digit turbine ever. Not to be left behind, GE Renewable Energy has developed a 14-megawatt machine with a 260-meter tower and 107-meter blades, also rolling out by 2021.

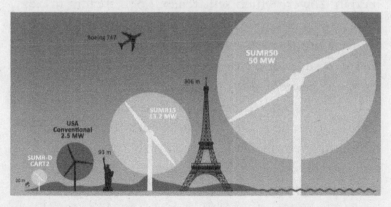

Comparisons of wind turbine heights and blade diameters

That is clearly pushing the envelope, although it must be noted that still-larger designs have been considered. In 2011, the UpWind project released what it called a "predesign" of a 20-megawatt offshore machine with a rotor diameter of 252 meters (three times the wingspan of an Airbus A380) and a hub diameter of 6 meters. So far, the limit of the largest conceptual designs stands at 50 megawatts, with a height exceeding 300 meters and 200-meter blades that could flex (much like palm fronds) in furious winds.

To imply, as an enthusiastic promoter did, that building such a structure would pose no fundamental technical problems because it stands no higher than the Eiffel Tower, built over 130 years ago, is to choose an inappropriate comparison. If the constructible height of an artifact were the determinant of wind turbine design, then we might as well refer to the Burj Khalifa in Dubai, a skyscraper that topped 800 meters in 2010, or to the Jeddah Tower, which will reach 1,000 meters in 2021. Erecting a tall tower is no great problem; it's quite another proposition, however, to engineer a tall tower that can support a massive nacelle and rotating blades for many years of safe operation.

Larger turbines must face the inescapable effects of scaling. Turbine power increases with the square of the radius swept by its blades: a turbine with blades twice as long would, theoretically, be four times as powerful. But the expansion of the surface swept by the rotor puts a greater strain on the entire assembly, and because blade mass should (at first glance) increase as a cube of blade length, larger designs should be extraordinarily heavy. In reality, designs using lightweight synthetic materials and balsa can keep the actual exponent to as little as 2.3.

Even so, the mass (and hence the cost) adds up. Each of the three blades of Vestas' 10-megawatt machine will weigh 35 tons, and the nacelle will come to nearly 400 tons (for the latter mass, think of lifting six Abrams main battle tanks a few hundred meters aloft). GE's record-breaking design will have blades of 55 tons, a nacelle of 600 tons, and a tower of 2,550 tons. Merely transporting such long and massive blades is an unusual challenge, although it could be made easier by using a segmented design.

Exploring likely limits of commercial capacity is more useful than forecasting specific maxima for given dates. Available wind turbine power is equal to half the density of the air (which is 1.23 kilograms per cubic meter) times the area swept by the blades (pi times the radius squared) times the cube of wind velocity. Assuming a wind velocity of 12 meters per second and an energy conversion coefficient of 0.4, then a 100-megawatt turbine would require rotors nearly 550 meters in diameter.

To predict when we'll get such a machine, just answer this question: When will we be able to produce 275-meter blades of plastic composites and balsa, figure out their transport and their coupling to nacelles hanging 300 meters above the ground, ensure their survival in cyclonic winds, and guarantee their reliable operation for at least 15 or 20 years? Not soon.

The slow rise of photovoltaics

In March 1958, a rocket lifted off from Cape Canaveral bearing the Vanguard 1 satellite: a small, 1.46-kilogram aluminum sphere that was the first to use photovoltaic (PV) cells in orbit.

As a safeguard, one of the satellite's two transmitters drew power from mercury batteries, but they failed after just three months. Thanks to the photoelectric effect, the six small monocrystalline silicon cells—absorbing light (photons) at the atomic level and releasing electrons—could deliver a total of just 1 watt, and it kept powering a beacon transmitter until May 1964.

It happened because, in space, cost was no object. In the mid-1950s, PV cells ran about $300 per watt. The cost fell to about $80/W in the mid-1970s, to $10/W by the late 1980s, to $1/W by 2011, and by late 2019 PV cells were selling for just 8–12 cents per watt, with further declines certain to come (of course, the cost of installing PV panels and associated equipment in order to generate electricity is substantially higher, depending on the scale of a project: they now range from tiny roof installations to large solar fields in deserts).

This is good news, because PV cells have a higher power density than any other form of renewable energy

An aerial view of the Ouarzazate Noor Power Station in Morocco.
At 510 MW, it is the world's largest central solar
power and photovoltaic installation

conversion. Even as an annual average they already reach 10 watts per square meter in sunny places, more than an order of magnitude higher than biofuels can manage. And with rising conversion efficiencies and better tracking, it should be possible to increase the annual capacity factors by 20–40 percent.

But it has taken quite a while to get to this point. Edmond Becquerel first described the photovoltaic effect in a solution in 1839, and William Adams and Richard Day discovered it in selenium in 1876. Commercial opportunities opened up only when the silicon cell was invented at Bell Telephone Laboratories in 1954. Even then, the cost per watt remained around $300 (more than $2,300 in 2020 money), and except for use in a few toys, PVs were just not practical.

It was Hans Ziegler, an electronics engineer with the US Army, who overcame the US Navy's initial decision to use only batteries on the Vanguard. During the 1960s, PV cells made it possible to power much larger satellites that revolutionized telecommunications, spying from space, weather forecasting, and the monitoring of eco-systems. As costs declined, applications multiplied, and PV cells began to power lights in lighthouses, offshore oil and gas drilling rigs, and railway crossings.

I bought my first solar scientific calculator—the Texas Instruments TI-35 Galaxy Solar—when it was introduced in 1985. Its four cells (each one about 170 square millimeters) still serve me well, more than 30 years later.

But serious PV electricity generation had to wait for further module price declines. By 2000, global PV generation supplied less than 0.01 percent of global electricity; a decade later, the share rose by an order of magnitude to 0.16 percent; and by 2018 it stood at 2.2 percent, still a small fraction compared to the share of electricity produced by the world's hydro stations (nearly 16 percent in 2018). In some sunny regions, solar generation is now making a discernible difference, but in global terms it still has a way to go before it rivals falling water.

Not even the most optimistic forecast—that of the International Renewable Energy Agency—expects PV output to close that gap by 2030. But PV cells might be generating 10 percent of the world's electricity by 2030. By that time, some seven decades will have passed since Vanguard 1's small cells began to power its beacon transmitter, and some 150 years since the photovoltaic effect was first discovered in a solid. Energy transitions on a global scale take time.

Why sunlight is still best

You can roughly track the advance of civilization by the state of its lighting—above all, its power, cost, and luminous efficacy. That last measure refers to the ability of a light source to produce a meaningful response in the eye, and it is the total luminous flux (in lumens) divided by the rated power (in watts).

Under photopic conditions (that is, under bright light, which allows color perception), the luminous efficacy of visible light peaks at 683 lm/W at a wavelength of 555 nanometers. That's in the green part of the spectrum— the color that seems, at any given level of power, to be the brightest.

For millennia, our sources of artificial light lagged three orders of magnitude behind this theoretical peak. Candles had a luminous efficacy of just 0.2 to 0.3 lm/W; coal gas lights (common in European cities during the 19th century) did five or six times as well; and the carbon filaments of Edison's early bulbs hardly did better than that. Efficacies took a leap with metal filaments, first with osmium in 1898 to 5.5 lm/W, then after 1901 with tantalum to 7 lm/W, and a decade or so later tungsten radiating in a vacuum got up to 10 lm/W. Putting a tungsten filament in a mixture of nitrogen and argon raised

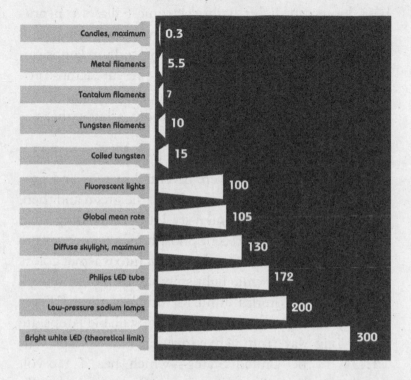

Lumens per watt

Candles, maximum	0.3
Metal filaments	5.5
Tantalum filaments	7
Tungsten filaments	10
Coiled tungsten	15
Fluorescent lights	100
Global mean rate	105
Diffuse skylight, maximum	130
Philips LED tube	172
Low-pressure sodium lamps	200
Bright white LED (theoretical limit)	300

the efficacy of common household lamps to 12 lm/W, and coiling the filaments, beginning in 1934, helped to bring incandescent efficacies to more than 15 lm/W for 100-watt lamps, which were the standard source of bright light during the first two post–Second World War decades.

Lights based on different principles—low-pressure sodium lamps and low-pressure mercury vapor lamps (fluorescent lights)—were introduced during the 1930s, but they only came to widespread use in the 1950s. Today's best fluorescent lights with electronic ballasts can produce about 100 lm/W; high-pressure sodium lamps put out up to 150 lm/W; and low-pressure sodium lamps can reach 200 lm/W. However, the low-pressure lamps produce only monochromatic yellow light at 589 nanometers, which is why they aren't used in homes but rather for street illumination.

Our best hope now rests on light-emitting diodes (LEDs). The first ones were invented in 1962 and provided only red light; a decade later came green and then, during the 1990s, high-intensity blue. By coating such blue LEDs with fluorescent phosphors, engineers were able to convert some of the blue light into warmer colors and thus produce white light suitable for indoor illumination. The theoretical limit for bright white LEDs is about 300 lm/W, but commercially available lamps are still a long way from achieving that rate. Philips sells LEDs in the United States—which has a 120-volt standard—that offer a luminous efficacy of 89 lm/W, for 18-watt soft white and dimmable bulbs (replacing 100-watt incandescent lights). In Europe, where the voltage ranges from 220 to 240 volts, the company sells a 172 lm/W LED tube (replacing the 1.5-meter-long European fluorescent tubes).

High-efficacy LEDs are already delivering significant electricity savings worldwide—it also helps that they can provide light for three hours a day for about 20 years, and if you forget to switch them off you will hardly notice on your next electricity bill. But, much like all other sources of artificial light, they still cannot match natural light's spectrum. Incandescent lights gave out too little blue light, and fluorescent lights had hardly any red; LEDs have too little intensity in the red part of the spectrum and too much in the blue part. They don't quite please the eye.

Light efficacies of artificial sources have improved by two orders of magnitude since 1880—but replicating sunlight indoors still remains beyond our reach.

Why we need bigger batteries

It would be a lot easier to expand our use of solar and wind energy if we had better ways to store the large quantities of electricity we'd need to cover gaps in the flow of that energy.

Even in sunny Los Angeles, a typical house roofed with enough photovoltaic panels to meet its average needs would still face daily shortfalls of up to 80 percent of the demand in January and daily surpluses of up to 65 percent in May. You can take such a house off the grid only by installing a voluminous and expensive assembly of lithium-ion batteries. And even a small national grid—one handling 10 to 30 gigawatts—could rely entirely on intermittent sources only if it had gigawatt-scale storage capable of working for many hours.

Since 2007, more than half of humanity has lived in urban areas, and by 2050 more than 6.3 billion people will live in cities, accounting for two-thirds of the global population, with a rising share in megacities of more than 10 million people (see THE RISE OF MEGA-CITIES, p. 44). Most of those people will live in high-rises, so there will be only a limited possibility of local generation, but they'll need an unceasing supply

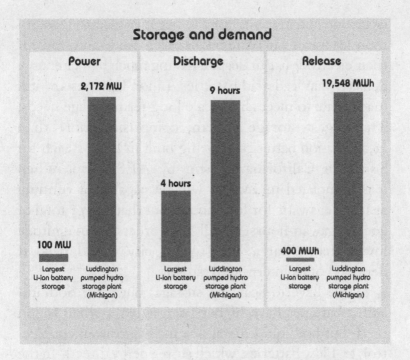

Storage and demand

Power

2,172 MW

100 MW

Largest
Li-ion battery
storage

Luddington
pumped hydro
storage plant
(Michigan)

Discharge

9 hours

4 hours

Largest
Li-ion battery
storage

Luddington
pumped hydro
storage plant
(Michigan)

Release

19,548 MWh

400 MWh

Largest
Li-ion battery
storage

Luddington
pumped hydro
storage plant
(Michigan)

of electricity to power their homes, services, industries, and transportation.

Think about an Asian megacity hit by a typhoon for a day or two. Even if long-distance lines could supply more than half of the city's demand, it would still need many gigawatt-hours from storage to tide it over until intermittent generation could be restored (or use fossil fuel backup—the very thing we're trying to get away from).

Lithium-ion (Li-ion) batteries are today's storage workhorses in both stationary and mobile applications. They deploy a lithium compound for their positive electrode and graphite for the negative electrode (common lead-acid car batteries use lead oxide and lead for their electrodes). But despite having much higher energy density than lead-acid batteries, Li-ion batteries are still inadequate to meet large-scale long-term storage needs. The largest storage system, comprising more than 18,000 Li-ion batteries, is being built in Long Beach for Southern California Edison by AES Corp. When it is completed in 2021, it will be capable of running at 100 megawatts for four hours. But that energy total of 400 megawatt-hours is still two orders of magnitude lower than what a large Asian city would need if deprived of its intermittent supply.

So we have to scale up storage, but how? Sodium-sulfur batteries have higher energy density than Li-ion ones, but hot liquid metal is a most inconvenient electrolyte. Flow batteries, which store energy directly in the electrolyte, are still in an early stage of deployment. Supercapacitors can't provide electricity over a long enough time. And compressed air and flywheels, the perennial favorites of popular journalism, have made it into only a dozen or so small or midsize installations. Perhaps the best long-term hope is to utilize cheap solar electricity to decompose water by electrolysis and use the produced hydrogen as a multipurpose fuel, but such a hydrogen-based economy is not imminent.

And so, when going big, we must still rely on a technology introduced in the 1890s: pumped storage. You build one reservoir high up, link it with pipes to another one lower down, and use cheaper, nighttime electricity to pump water uphill so that it can turn turbines during times of peak demand. Pumped storage accounts for more than 99 percent of the world's storage capacity, but inevitably it entails energy loss on the order of 25 percent. Many installations have short-term capacities in excess of 1 gigawatt—the largest one is about 3 gigawatts—and more than one would be needed for a megacity completely dependent on solar and wind generation.

But most megacities are nowhere near the steep escarpments or deep-cut mountain valleys you'd need for pumped storage. Many—including Shanghai, Kolkata, and Karachi—are on coastal plains. They could rely on pumped storage only if it were provided through long-distance transmission.

The need for more compact, more flexible, larger-scale, less costly electricity storage is self-evident. But the miracle has been slow in coming.

Why electric container
ships are a hard sail

Just about everything you wear or use around the house once sat in steel boxes on ships whose diesel engines propelled them from Asia, emitting particulates and carbon dioxide. Surely, you would think, we can do better.

After all, we've had electric locomotives for more than a century and high-speed electric trains for more than half a century, and recently we have been expanding the global fleet of electric cars. Why not get electric container ships?

Actually, the first one is scheduled to operate in 2021: the *Yara Birkeland*, built by Marin Teknikk in Norway, is not only the world's first electric-powered, zero-emissions container ship, but also the first autonomous commercial vessel.

But don't write off giant diesel-powered container ships and their critical role in a globalized economy just yet. Here is a back-of-the-envelope calculation that explains why . . .

Containers come in different sizes, but most are the standard twenty-foot equivalent units (TEU)— rectangular prisms 6.1 meters (20 feet) long and 2.4 meters wide. The first small container ships of the 1960s carried mere hundreds of TEUs; now four ships

Model of *Yara Birkeland*

launched in 2019 and belonging to MSC Switzerland (*Gülsün*, *Samar*, *Leni* and *Mia*) hold the record, at 23,756 TEUs each. When they travel very slowly (16 knots, in order to save fuel) these ships can make the journey from Hong Kong to Hamburg (via the Suez Canal)—more than 21,000 kilometers—in 30 days.

Now consider the *Yara Birkeland*. It will carry just 120 TEUs, its service speed will be 6 knots, and its longest intended operation will be 30 nautical miles—between Herøya and Larvik, in Norway. Today's state-of-the-art diesel container vessels thus carry nearly 200 times as many boxes over distances almost 400 times as long, at speeds three to four times as fast as the pioneering electric ship can handle.

What would it take to make an electric ship that can carry up to 18,000 TEUs, now a common intercontinental load? In a 31-day trip, most of today's efficient diesel vessels burn 4,650 tons of fuel (low-quality residual oil or diesel), with each ton packing 42 gigajoules. That's an energy density of about 11,700 watt-hours per kilogram, versus 300 Wh/kg for today's lithium-ion batteries—a nearly 40-fold difference.

The total fuel demand for the trip is about 195 terajoules, or 54 gigawatt-hours. Large diesel engines (and those installed in container ships are the largest we have) are about 50 percent efficient, which means that the energy actually used for propulsion is half the total fuel demand, or about 27 gigawatt-hours. To match that demand, large electric motors operating at 90 percent efficiency would need about 30 gigawatt-hours of electricity.

Load the ship with today's best commercial Li-ion batteries (300 Wh/kg), and still it would have to carry about 100,000 tons of them to go nonstop from Asia to Europe in a month (for comparison, electric cars contain about 500 kilograms, or 0.5 tons, of Li-ion batteries). Those batteries alone would take up about 40 percent of maximum cargo capacity—an economically ruinous proposition, never mind the difficulties involved in charging and operating the ship. And even if we push batteries to an energy density of 500 Wh/kg sooner than might be expected, an 18,000-TEU vessel would still need nearly 60,000 tons of them for a long intercontinental voyage at a relatively slow speed.

The conclusion is obvious. To have an electric ship whose batteries and motors weighed no more than the fuel (about 5,000 tons) and the diesel engine (about 2,000 tons) in today's large container vessels, we would need batteries with an energy density more than 10 times as high as today's best Li-ion units.

But that's a tall order indeed: in the past 70 years, the energy density of the best commercial batteries hasn't even quadrupled.

The real cost of electricity

In many affluent countries, the new century brought a shift in the long-term trajectory of electricity prices: they have increased not only in the monies of the day but even after adjustments for inflation. Even so, electricity remains an admirable bargain—though, as expected, a bargain with many national peculiarities, resulting not only from specific contribution by different sources but also from persistent governmental regulation.

Historical perspective shows the trajectory of an extraordinary value, and that explains electricity's ubiquity in the modern world. When adjusted for inflation (and expressed in constant 2019 monies), the average price of US residential electricity fell from $4.81 per kilowatt-hour in 1902 (the first year for which the national mean is available) to 30.5 cents in 1950, then to 12.2 cents in 2000; and in early 2019 it was just marginally higher, at 12.7 cents/kWh. This represents a relative decline of more than 97 percent—or, stated in reverse, a dollar now buys nearly 38 times more electricity than it did in 1902. But, during that period, average (again, inflation-adjusted) manufacturing wages nearly sextupled, which means that in a blue-collar household, electricity is now more than 200 times more affordable (its effective

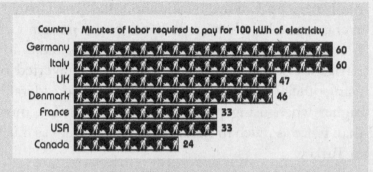

Country	Minutes of labor required to pay for 100 kWh of electricity	
Germany		60
Italy		60
UK		47
Denmark		46
France		33
USA		33
Canada		24

earning-adjusted cost is less than 0.5 percent of the 1902 rate) than it was almost 120 years ago.

But we buy electricity in order for it to be converted into light or kinetic energy or heat, and the intervening efficiency improvements have made its final uses an even greater bargain, with lighting offering the most impressive gain. In 1902, a lightbulb with a tantalum filament produced 7 lumens per watt; in 2019, a dimmable LED light delivers 89 lm/W. That means that a lumen of electric light for a working-class household is now approximately 2,500 times more affordable than it was in the early 20th century.

International perspective shows some surprising differences. The US has cheaper residential electricity than any affluent nation with the exception of Canada and Norway, the high-income countries with the highest shares of hydroelectricity generation (59 and 95 percent, respectively). When using prevailing exchange rates, the US residential price is about 55 percent of the European Union's mean, about half the Japanese average, and less than 40 percent of the German rate. Electricity prices in India, Mexico, Turkey, and South Africa are lower than in the US when converted by using official exchange rates, but they are considerably higher when using purchasing power parities: more than twice as much in India; nearly three times as much in Turkey.

When reading reports about dramatically falling costs of photovoltaic cells (see THE SLOW RISE OF PHOTO-VOLTAICS, p. 154) and the highly competitive pricing of wind turbines, a naive observer might conclude that the rising shares of new renewables (solar and wind) will usher in an era of falling electricity prices. But in reality, the opposite has been true. Before the year 2000, when the country embarked on its large-scale and expensive program of expanded renewable electricity generation (*Energiewende*), German residential electricity prices were low and declining—bottoming out at less than €0.14/kWh in the year 2000.

By 2015, Germany's combined solar and wind capacity of nearly 84 gigawatts had surpassed the total installed in fossil fuel plants, and by March 2019 more than 20 percent of all electricity came from the new renewables—but electricity prices had more than doubled in 18 years, to €0.29/kWh. The EU's largest economy thus has the continent's second-highest electricity prices: only in heavily wind-dependent Denmark (in 2018, 41 percent of its generation was from wind) is the price higher, at €0.31/kWh. A similar contrast can be seen in the US. In California, with its increasing share of new renewables, electricity prices have been rising five times faster than the national mean, and are now nearly 60 percent higher than the countrywide average.

The inevitably slow pace of energy transitions

In 1800, only the UK and a few localities in Europe and North China burned coal to generate heat—98 percent of the world's primary energy came from biomass fuels, mostly from wood and charcoal; in deforested regions energy also came from straw and from dried animal dung. By 1900, as coal mining expanded and oil and gas production began in North America and Russia, biomass supplied half of the world's primary energy; by 1950 it was still nearly 30 percent, and at the beginning of the 21st century it had declined to 12 percent, though in many sub-Saharan countries it remains above 80 percent. Clearly, it has taken a while to accomplish the transition from new carbon (in plant tissues) to old (fossil) carbon in coal, crude oil, and natural gas.

We are now in the earliest stages of a much more challenging transition: the decarbonization of the global energy supply which is needed in order to avoid the worst consequences of global warming. Contrary to a common impression, this transition has not been proceeding at a pace resembling the adoption of cellphones. In absolute terms, the world has been running into—not away from—carbon (see RUNNING INTO

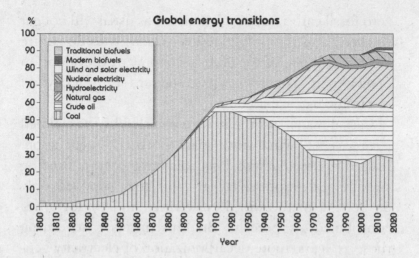

%

Global energy transitions

Legend:
- Traditional biofuels
- Modern biofuels
- Wind and solar electricity
- Nuclear electricity
- Hydroelectricity
- Natural gas
- Crude oil
- Coal

Year

CARBON, p. 303), and in relative terms our gains remain in single digits.

The first United Nations Framework Convention on Climate Change was held in 1992. In that year, fossil fuels (using the conversion of fuels and electricity to a common denominator favored by BP in its annual statistical report) provided 86.6 percent of the world's primary energy. By 2017, they supplied 85.1 percent, a reduction of a mere 1.5 percent in 25 years.

This key indicator of the global energy transition pace is perhaps the most convincing reminder of the world's continued fundamental dependence on fossil carbon. Can a marginal slip of 1.5 percent in a quarter-century be followed in the coming 25–30 years with the substitution of some 80 percent of the world's primary energy for non-carbon alternatives, in order to come close to zero fossil carbon by 2050? Business as usual will not get us there, and the only plausible scenarios to do so are either a collapse of the global economy or the adoption of new energy sources at a pace and on a scale far beyond our immediate capabilities.

Casual readers of the news get misled by the claimed advances of wind and solar electricity generation. Indeed, these renewable sources have been advancing steadily and impressively: in 1992 they supplied only 0.5 percent of the world's electricity, and by 2017 they contributed 4.5 percent. But this means that, during those 25 years, more decarbonization of electricity generation was due to expanded hydroelectricity generation than to combined solar and wind installations. And because only about 27 percent of the world's final energy consumption is electricity, these advances translate to a much smaller share of overall carbon reduction.

But solar and wind electricity generation are now mature industries, and new capacities can be added quickly—increasing the pace of decarbonizing the electricity supply. In contrast, several key economic sectors depend heavily on fossil fuels and we do not have any non-carbon alternatives that could replace them rapidly and on the requisite massive scales. These sectors include long-distance transportation (now almost totally reliant on aviation kerosene for jetliners, and diesel, bunker fuel, and liquefied natural gas for container, bulk, and tanker vessels); the production of more than a billion tons of primary iron (requiring coke made from coal for smelting iron ores in blast furnaces) and more than 4 billion tons of cement (made in massive rotating kilns fired by low-quality fossil fuels); the synthesis of nearly 200 million tons of ammonia and some 300 million tons of plastics (starting with compounds derived from natural gas and crude oil); and space heating (now dominated by natural gas).

These realities, rather than any wishful thinking, must guide our understanding of primary energy transitions. Displacing 10 billion tons of fossil carbon is a fundamentally different challenge than ramping up the sales of small portable electronic devices to more than a billion units a year; the latter feat was achieved in a matter of years, the former one is a task for many decades.

TRANSPORT
How We Get Around

Shrinking the journey across the Atlantic

Commercial sailing ships had long taken three—sometimes four—weeks to make the eastbound crossing of the Atlantic; the westbound route, against the wind, usually took six weeks. The first steamship to make the eastward crossing was in 1833, when the Quebec-built SS *Royal William* went to England after stopping to take on coal in Nova Scotia. It was only in April 1838 that steamships pioneered the westward route. And it happened in an unexpectedly dramatic way.

Isambard Kingdom Brunel, one of the great 19th-century British engineers, built the SS *Great Western* for the Great Western Steamship Company's planned Bristol–New York run. The ship was ready on March 31, 1838, but a fire onboard delayed the scheduled departure until April 8.

Meanwhile, the British and American Steam Navigation Company tried to steal a march by chartering the SS *Sirius*, a small wooden paddle-wheel vessel built for the Irish (London–Cork) service. The *Sirius* left Cobh, Ireland, on April 4, 1838, its boilers operating under 34 kilopascals for a peak engine power of 370 kilowatts (for comparison, a 2019 Ford Mustang is rated at 342 kilowatts). With 460 tons of coal onboard, the ship could

Brunel's *Great Western*: the paddle-wheel, steam-engine-powered
ship still had the rigging for sails

travel nearly 5,400 kilometers (2,916 nautical miles)—
almost, but not quite, all the way to New York Harbor.

In contrast, the *Great Western* was the world's largest
passenger ship, with 128 beds in first class. The ship's
boilers also worked at 34 kilopascals but its engines
could deliver about 560 kilowatts (the power of today's
industrial diesel generators), and on its first transatlantic
journey it averaged 16.04 kilometers per hour (slower
than today's best marathon runners, who average just
over 21 km/h). Even with its four-day head start, the
Sirius (averaging 14.87 km/h) barely beat the larger and
faster ship, arriving in New York on April 22, 1838—
after 18 days, 14 hours, and 22 minutes.

Later stories dramatized the final dash by claiming that the *Sirius* ran out of coal and had to burn furniture and even its spars to reach the port. Not true, but it did have to burn several drums of resin. When the *Great Western* arrived the next day, after 15 days and 12 hours, even after burning 655 tons of coal it still had 200 tons to spare.

Steam more than halved the transatlantic travel time, and new records kept coming. By 1848, Cunard's SS *Europa* crossed in 8 days and 23 hours. By 1888, a crossing took barely over 6 days; and in 1907, the steam turbine–powered RMS *Lusitania* won the Blue Riband (the trophy for the fastest Atlantic crossing) with a time of 4 days, 19 hours, and 52 minutes. The final record-holder, SS *United States*, made it in 3 days, 10 hours, and 40 minutes, in 1952.

The next era, in which commercial piston-engine aircraft crossed in 14 or more hours, was a brief one, because by 1958 America's first commercial turbojet, the Boeing 707, was making regularly scheduled flights from London to New York in less than 8 hours (see WHEN DID THE JET AGE BEGIN?, p. 204). Cruising speeds have not changed much: the Boeing 787 Dreamliner cruises at 913 km/h, and London–New York flights still last about 7.5 hours.

The expensive, noisy, and ill-fated supersonic Concorde could do it in 3.5 hours, but that bird will never fly again. Several companies are now developing supersonic transport planes, and Airbus has patented a hypersonic concept with a cruising speed of 4.5 times the speed of sound. Such a plane would arrive at JFK International just one hour after leaving Heathrow.

But do we really need such speed at a much higher energy cost? Compared with the *Sirius*'s time in 1838, we have cut the crossing time by more than 98 percent. The time aloft is just right for reading a substantial novel— or even this book, perhaps.

Engines are older than bicycles!

Some technical advances are delayed by either a failure of imagination or a concatenation of obstructive cir- cumstances. I can think of no better example of both of these than the bicycle.

Over two centuries ago in Mannheim, on June 12, 1817, Karl Drais, a forester in Germany's grand duchy of Baden, demonstrated for the first time his *Laufmaschine* ("running machine"), later also known as a "draisine" or hobby-horse. With the seat in the middle, front-wheel steering, and wheels of the same diameter, it was the archetype of all later vehicles that required constant bal- ancing. However, it was propelled not by pedaling but by pushing one's feet against the ground, Fred Flint- stone fashion.

Drais covered nearly 16 kilometers in little more than an hour on his heavy wooden bicycle, faster than the typical horse-drawn carriage. But it's obvious, today at least, that the design was clumsy and that there weren't yet enough suitable hard-top roads. Why, in the dec- ades after 1820 that abounded with such inventions as locomotives, steamships, and manufacturing tech- niques, did it take so long to come up with a means of

John Kemp Starley's Rover safety bicycle

propulsion that could make the bicycle a practical machine, able to be ridden by anybody but infants?

Several answers are obvious. Wooden bicycles were heavy and clumsy, and the cheap steel parts (frame, rims, spokes) required to design durable machines were not yet available. Unpaved roads offered only uncomfortable rides. Pneumatic tires were not invented until the late 1880s (see the next chapter). And urban incomes had to rise first in order to allow for a larger-scale adoption of what was essentially a leisure machine.

Only in 1866 did Pierre Lallement get his US patent for a bicycle propelled by pedals attached to a slightly larger front wheel. And starting in 1868, Pierre Michaux made this *vélocipède* design popular in France. But the Michaudine did not become the precursor of modern bicycles; it was just an ephemeral novelty. The entire 1870s and the early 1880s were dominated by high-wheelers (also known as "ordinary" or penny-farthing bicycles), their pedals attached directly to the axles of front wheels with diameters of up to 1.5 meters to provide a longer distance per pedal revolution. These clumsy machines could be fast, but they were also difficult to mount and tricky to steer; their use called for dexterity, stamina, and a tolerance for dangerous falls.

It wasn't until 1885 that two British inventors, John Kemp Starley and William Sutton, began to offer their Rover safety bicycles with equally sized wheels, direct steering, a chain and sprocket drive, and a tubular steel frame. Although it was not quite yet in the classic diamond shape, it was a truly modern bicycle design, ready for mass-adoption. The trend accelerated in 1888, with the introduction of John Dunlop's pneumatic tires.

So, a simple balancing machine consisting of two equally sized wheels, a minimal metal frame, and a short drive chain emerged more than a century *after* Watt's improved steam engines (1765), more than half a century *after* the introduction of mechanically far more complex locomotives (1829), years *after* the first commercial generation of electricity (1882)—but *concurrently* with the first designs of automobiles. The first light internal combustion engines were mounted on three- or four-wheel carriages by Karl Benz, Gottlieb Daimler, and Wilhelm Maybach in 1886.

And although cars changed enormously between 1886 and 1976, bicycle design remained remarkably conservative. The first purpose-built mountain bikes came only in 1977. Widespread adoption of such novelties as expensive alloys, composite materials, strange-looking frames, solid wheels, and upturned handlebars began only during the 1980s.

The surprising story of inflatable tires

Famous inventions are few, and they generally carry the name of a person or institution. Edison's lightbulb and Bell Labs' transistor are perhaps the most notable examples in this very small category, although Edison did not invent the lightbulb (just its more durable version), and Bell Labs merely reinvented the transistor (the solid-state device was patented in 1925 by Julius Edgar Lilienfeld).

At the other end of the recognition spectrum is the much larger category of epoch-making inventions whose origins are obscure. There is no better example of this than the inflatable tire, invented by one John Boyd Dunlop, a Scotsman living in Ireland. His British patent dates back over 130 years, to December 7, 1888.

Before Dunlop, the best bet was the solid rubber tire, which had been available ever since Charles Goodyear's vulcanization process (heating rubber with sulfur to increase its elasticity, patented in 1844) made it possible to produce durable rubber. Although such tires were a major improvement on solid wooden wheels or spoked wheels with iron rims, they still gave a jarring ride.

Dunlop devised his prototype, in 1887, to smooth the bumpy ride of his son's tricycle. It was a primitive

John Boyd Dunlop riding on his invention

product—simply an inflated tube that was tied off, wrapped in linen, and fastened to a solid wooden tricycle wheel by nails.

An improved version found immediate use among the growing numbers of enthusiastic bicycle riders, and a company was set up to manufacture the tires. However, as with so many other inventions, Dunlop's patent was eventually invalidated because it turned out that another Scotsman, Robert William Thomson, had previously patented the idea, even though he never made a practical product.

Still, Dunlop's invention stimulated work on larger

tires for the newly invented automobile. In 1885, Karl Benz's first three-wheeled Patent Motorwagen had solid rubber tires. Six years later, the Michelin brothers, André and Édouard, introduced their version of detachable rubber tires for bicycles, and in 1895 their two-seater, *L'Éclair*, became the first automobile with inflatable rubber tires to enter the nearly 1,200-kilometer Paris–Bordeaux–Paris race. Because its tires needed changing every 150 kilometers, *L'Éclair* ended up in ninth place.

It was a temporary setback. Sales did well, and Bibendum, the bulging-tire man, became Michelin's symbol in 1898. One year later the company's tires shod the torpedo-shaped *La Jamais Contente* ("The Never Satisfied"), a Belgian electric car that topped 100 kilometers per hour. In 1913, Michelin introduced the removable steel wheel and hence the convenience of having a spare wheel in the trunk—a setup that has endured to this day.

John Dunlop finally entered the Automotive Hall of Fame in 2005, and the Dunlop brand is still around, now owned by the Goodyear Tire and Rubber Company, the world's third-largest maker of tires. Japan's Bridgestone is the leader, but Michelin is a close second—the rare example of a company that has stayed near the top of its industry for more than a century.

Tires are quintessential products of the industrial age—heavy, bulky, polluting, still extremely difficult to dispose of—but even in our information era they are still needed in ever-higher numbers. Tire companies must meet the worldwide demand for nearly 100 million new road vehicles every year, and for replacements for the global fleet of more than 1.2 billion.

Dunlop would be astounded by what he began. So much for the much-hyped dematerialization of our world that artificial intelligence is supposed to have started.

When did the age of the car begin?

In 1908, Henry Ford had been working in the auto business for more than a decade, and the Ford Motor Company, five years old and already profitable, had so far followed its peers by catering to the well-to-do. Its Model K, introduced in 1906, was priced at around $2,800, and the smaller Model N, introduced the same year, sold for $500—about what the average person earned in a year.

Then, on August 12, 1908, the age of the automobile began, because on that day the first Ford Model T was assembled at Detroit's Piquette Avenue Plant. It went on sale on October 1.

Ford made his goals clear: "I will build a car for the great multitude. It will be large enough for the family, but small enough for the individual to run and care for. It will be constructed of the best materials . . . after the simplest designs that modern engineering can devise. But it will be so low in price that no man making a good salary will be unable to own one." He met those goals, thanks to his vision and to the talent he was able to recruit, notably the designers Childe Harold Wills, Joseph A. Galamb, Eugene Farkas, Henry Love, C.J. Smith, Gus Degner, and Peter E. Martin.

Ford Model T

The four-cylinder water-cooled engine put out 15 kilo-watts (today's small cars are commonly eight times more powerful), the top speed was 72 kilometers per hour, and the price was low. The Runabout, the most popular model, sold for $825 in 1909, but continuous design and manufacturing improvements let Ford lower the price to $260 by 1925. That represented about two and a half months' wages for the average worker at the time. Today, the average new-car price in the United States is $34,000, or about 10 months' median salary. In the UK, popular small car models average about £15,000 (about $20,000).

Introduction of a moving assembly line at Detroit's

Highland Park factory in 1913 brought substantial economies of scale: by 1914 the plant was already turning out 1,000 automobiles a day. And Ford's decision to pay unprecedented wages for unskilled assembly labor assured uninterrupted production. In 1914 the rate was more than doubled, to $5 a day, and the working day was reduced to eight hours.

The outcome was impressive. The Ford Motor Company produced 15 percent of all US cars in 1908, 48 percent in 1914, and 57 percent in 1923. By May 1927, when the production run ended, the company had sold 15 million Model Ts.

Ford stood at the very beginning of manufacturing globalization, using standardized procedures and dispersing car assembly around the world. Foreign assembly began in Canada and then fanned out to the United Kingdom, Germany, France, Spain, Belgium, and Norway, as well as to Mexico, Brazil, and Japan.

But even though Ford staked much on this one car, it didn't quite become the bestselling vehicle in history. That primacy belongs to the "people's car" of Germany—the Volkswagen. Soon after he came to power, Adolf Hitler decreed its specifications, insisted on its distinctive beetle-like appearance, and ordered Ferdinand Porsche to design it.

By the time it was ready for production, in 1938, Hitler had other plans, and the car's assembly didn't begin until 1945, in the British-occupied zone. German production ended in 1977, but the original VW Beetle continued to be assembled in Brazil until 1996 and in Mexico until 2003. The last car, made in Puebla, was number 21,529,464.

But in many ways the Beetle was just an updated emulation of the Model T. There can never be any dispute over who mass-produced the first affordable passenger car.

Modern cars have a terrible weight-to-payload ratio

A century ago, the bestselling car in the United States, Ford's Model T, wrung a watt from every 12 grams of its internal combustion engine. Now, engines in best-selling American cars are getting a watt per gram— a 92 percent improvement. That is the one bit of happy news I am going to impart in this chapter.

Now for the bad news: the US data show that during the past 100 years, average engine power has increased more than 11-fold, to about 170 kilowatts. This means that despite a huge drop of mass/power density, today's typical car engine is hardly lighter than it was a century ago—and the average car itself has become much heavier:

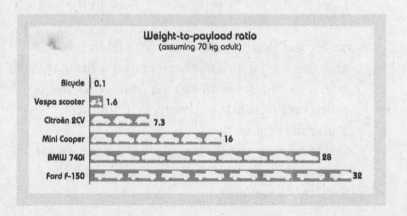

Weight-to-payload ratio
(assuming 70 kg adult)

Bicycle	0.1
Vespa scooter	1.6
Citroën 2CV	7.3
Mini Cooper	16
BMW 740i	28
Ford F-150	32

its mass has roughly tripled, reaching more than 1,800 kilograms (the average for all light-duty vehicles, nearly half
of which are pickups, SUVs, and minivans).

And because nearly three-quarters of US commuters
drive alone, you get the worst possible ratio of vehicle-
to-passenger weight.

That ratio is what matters. Because, for all the auto
industry's talk about "lightweighting"—using aluminum,
magnesium, and even carbon fiber–reinforced polymers
to reduce total weight—this ratio ultimately limits the
energy efficiency.

Here, in ascending order, are a few of the weight
ratios that a 70-kilogram passenger can achieve:

- 0.1 for a 7-kilogram bicycle
- 1.6 for Italy's 110-kilogram Vespa scooter
- 5 or less for a modern bus, and that's just if you
  count sitting passengers
- 7.3 for France's 510-kilogram Citroën 2CV
  (*deux chevaux*, or "two horses"), back in the
  1950s
- 7.7 for the Ford Model T introduced in 1908
  and also for Japan's *shinkansen* rapid train that
  began to run in October 1964 (the train's frugal
  ratio owes as much to design as it does to a
  high ridership rate)
- 12 for a Smart car, 16 for a Mini Cooper, 18 for
  my own Honda Civic LX, 20 and change for
  the Toyota Camry

- 26 for the average American light-duty vehicle in 2013
- 28 for the BMW 740i
- 32 for the Ford F-150, the bestselling American vehicle
- 39 for the Cadillac Escalade EXT.

Of course, you can get quite spectacular ratios by pairing the right car with the right driver. I regularly see a woman driving a Hummer H2 that easily weighs 50 times as much as she does. That's like going after a fly with a steam shovel.

To put it all in perspective, consider that the latest Boeing, the 787-10, does better than a small Citroën. Its maximum takeoff weight is 254 tons; with 330 passengers weighing 23 tons and another 25 tons of cargo, the overall weight-to-payload ratio comes to just 5.3.

Cars got heavy because part of the world got rich and drivers got coddled. Light-duty vehicles are larger, and they come equipped with more features, including automatic transmissions, air conditioning, entertainment and communication systems, and an increasing number of servomotors powering windows, mirrors, and adjustable seats. And new battery-heavy hybrid drives and electric cars will not be lighter: the small all-electric Ford Focus weighs 1.7 tons, General Motors' Volt is more than 1.7 tons, and the Tesla is just above 2.1 tons.

Lighter designs would help, but obviously nothing could halve (or quarter) the ratio as easily as having two or four people in a car. And yet in the United States, that is the hardest thing to enforce. The 2019 *State of the American Commute* reported that nearly three-quarters of commuters drive alone to work. Car commutes are much less common in Europe (36 percent in the UK), and even rarer in urban Japan (just 14 percent)—but average car sizes have been increasing both in the EU and in Japan.

And so the outlook is for ever-better engines or electric motors in heavier vehicles used in a way that results in the worst weight-to-payload ratios for any mechanized means of personal transportation in history.

These cars may be, by some definition, smart—but they are not wise.

Why electric cars aren't as great as we think (yet)

Let me begin with a disclaimer: I am neither promoting electric vehicles (EVs) nor denigrating them. I simply observe that the rational case for accepting them has been undermined by unrealistic market forecasts and a disregard for the environmental effects involved in producing and operating such vehicles.

Unrealistic forecasts have been, and continue to be, the norm. In 2010, Deutsche Bank predicted that EVs would claim 11 percent of the global market by 2020— in reality, they will be less than 4 percent. And this triumph of hope over experience continues. Recent forecasts for 2030 have EVs adding up to as much as 20 percent of the global car fleet, or as little as 2 percent. Bloomberg New Energy Finance sees 548 million EVs on roads by 2040; Exxon only 162 million.

EV enthusiasts have also neglected to note the environmental consequences of mass-scale conversion to electric drive. If EVs are to reduce carbon emissions (and thus minimize the extent of global warming), their batteries must not be charged with electricity generated from the combustion of fossil fuels. But in 2020, just over 60 percent of global electricity will originate in fossil fuels; about 12 percent will come from wind and

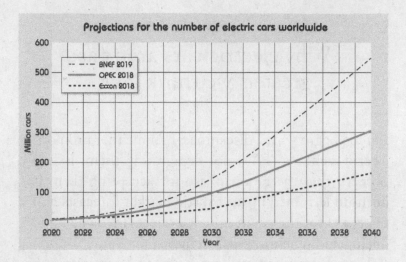

Projections for the number of electric cars worldwide

solar; and the rest from hydro energy and nuclear fission.

As a global mean, more than three-fifths of the electricity for an EV still comes from fossil carbon, but that fraction varies widely among countries and within them. EVs in my home province of Manitoba, Canada (where more than 99 percent of all electricity comes from large hydro stations) are clean hydro cars. Quebec, Canada (about 97 percent hydro) and Norway (about 95 percent hydro) come close to that. French EVs are largely nuclear-fission cars (the country gets some 75 percent of its electricity from fission). But in most of India (particularly Uttar Pradesh), China (particularly Shaanxi province), and Poland, EVs are overwhelmingly coal cars. The last thing we need is to push for the rapid introduction of a source of demand that would summon even more fossil fuel–based electricity generation.

And even if EVs all ran on renewable sources of electricity, greenhouse gases would still be emitted during the production of cement and steel for hydroelectric dams, wind turbines, and photovoltaic panels, and of course during the manufacture of the cars themselves (see WHAT'S WORSE FOR THE ENVIRONMENT—YOUR CAR OR YOUR PHONE?, p. 287).

EV production will have other environmental impacts, too. The Arthur D. Little management consultancy estimates that—based on a vehicle life of 20 years—the manufacture of an EV creates three times as much toxicity as that of a conventional vehicle. This is mostly due to the greater use of heavy metals. Similarly, a detailed comparative life-cycle analysis, published in the *Journal of Industrial Ecology*, found the production of EVs to involve substantially higher toxicity, both to human beings and to freshwater ecosystems.

I am not suggesting that these are arguments against the adoption of EVs. I am merely pointing out that the implications of the new technology must be appraised and understood before we accept any radical claims in its favor. We cannot simply imagine ideal, pollution-free machines and then will them into existence.

When did the jet age begin?

Dating the dawn of the jet age is hard because there were so many different "firsts." The first experimental takeoff of a jet-powered airplane was that of a warplane, the German Heinkel He 178, in August 1939 (fortunately, it entered the service too late to affect the outcome of the Second World War). The first flight of the first commercial design, the British de Havilland DH 106 Comet, was in July 1949, and its first British Overseas Airways Corporation commercial flight was in 1952. But four disasters (in October 1952 near Rome; in May 1953 in Calcutta; in January 1954, again near Rome; and in April 1954 near Naples) grounded the Comet fleet, and a redesigned airplane made the first transatlantic flight on October 4, 1958. Meanwhile, the Soviet Tupolev Tu-104 entered domestic service in September 1956.

But you can make a strong argument that the jet age began on October 26, 1958, when a Pan Am Boeing 707 took off from Idlewild Airport (now JFK International Airport) to Paris, on the first of its daily scheduled flights.

Several reasons justify that choice. The redesigned Comet was too small and unprofitable to begin a design dynasty, and there were no successor models. Meanwhile,

Sending off the first Boeing 707 flight

Tupolev's aircraft were used only by the countries of the Soviet bloc. The Boeing 707, however, inaugurated the industry's most successful design family, one that progressed relentlessly by adding another 10 models to its varied lineup.

The three-engine Boeing 727 was the first follow-on, in 1963; the four-engine 747, introduced in 1969, was perhaps the most revolutionary design in modern aviation; and the latest addition, the 787 Dreamliner series introduced in 2011, is made mostly of carbon fiber composites and is now able to fly on routes longer than 17 hours.

The 707 had a military pedigree: the plane started as a prototype of an in-air refueling tanker, and further development led to the KC-135A Stratotanker and finally to a four-engine passenger plane powered by small-diameter Pratt & Whitney turbojet engines, each with about 50 kilonewtons of thrust. By comparison, each of the two General Electric GEnx-1B high-bypass turbo-fan engines powering today's 787 delivers more than 300 kilonewtons at takeoff.

The first scheduled flight of the 707 *Clipper America* on October 26, 1958 was preceded by a welcoming ceremony, a speech by Juan Trippe (Pan Am's then-president), and a performance by a US Army band. The 111 passengers and 12 crew members had to make an unscheduled stop at Gander International Airport in Newfoundland, Canada, but even so they were able to land at Paris-Le Bourget Airport 8 hours and 41 minutes after leaving New York. By December the plane was flying the New York–Miami route, and in January 1959 it began to make the first transcontinental flights, from New York to Los Angeles.

Before the introduction of the wide-bodies—first the Boeing 747, and then the McDonnell Douglas DC-10 and the Lockheed L-1011 in 1970—Boeing 707s were the dominant long-distance jetliners. One of them brought me and my wife from Europe to the United States in 1969.

Gradual improvements in the Boeing family resulted in a vastly superior plane. In a standard two-class (business and economy) configuration, the first Dreamliner could seat about 100 more people than the 707-120, with a maximum takeoff weight nearly twice as great and a maximum range almost twice as long. Yet the Dreamliner consumes 70 percent less fuel per passenger-kilometer. And because it is built from carbon composites, the 787 can be pressurized to simulate a lower altitude than an aluminum fuselage will allow, resulting in greater comfort for passengers.

Eventually, Boeing made just over a thousand 707s. When Pan Am brought the plane out of retirement for a 25th-anniversary commemorative flight in 1983, it flew most of the original crew as passengers to Paris. But that was not the end of 707 service. A number of non-US airlines flew different models until the 1990s, and Iran's Saha Airlines did so as late as 2013.

Although today the 707 can be found only in jet junkyards, the plane's place in history remains secure. It represents the first effective and rewarding step in the evolution of commercial jet flight.

Why kerosene is king

Eliminating kerosene-based jet fuel will be one of the greatest challenges in creating a world without carbon emissions. Aviation accounts for only about 2 percent of the global volume of such emissions and some 12 percent of the total released by the transportation sector, but converting to electric drive is much harder for airplanes than for cars and trains.

Today's jet fuel—the most common formulation of which is called Jet A-1—has a number of advantages. It has a very high energy density, as it packs 42.8 megajoules into each kilogram (that is slightly less than gasoline but it can stay liquid down to −47°C), and it beats gasoline on cost, evaporative losses at high altitude, and risk of fire during handling. No real rivals yet exist. Batteries capacious enough for intercontinental flights carrying hundreds of people are still the stuff of science fiction, and we will not see wide-bodied planes fueled by liquid hydrogen anytime soon.

What we need is a fuel equivalent to kerosene that is derived from plant matter or organic waste. Such a bio-jet fuel would release no more CO_2 during combustion than plants sequester during growth. The proof of principle has been demonstrated: since 2007, test flights

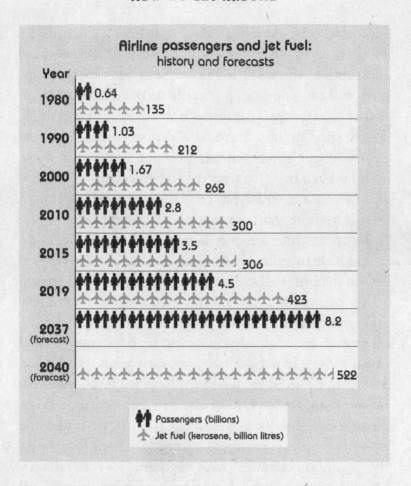

Airline passengers and jet fuel:
history and forecasts

Year	
1980	0.64 / 135
1990	1.03 / 212
2000	1.67 / 262
2010	2.8 / 300
2015	3.5 / 306
2019	4.5 / 423
2037 (forecast)	8.2
2040 (forecast)	522

Passengers (billions)

Jet fuel (kerosene, billion litres)

using blends of Jet A-1 and biojet have proved suitable as drop-in alternatives for modern aircraft.

In that time, some 150,000 flights have used blended fuel, but only five major airports have regular biofuel distribution (Oslo, Stavanger, Stockholm, Brisbane, and Los Angeles), with others offering occasional supply.

Biofuel use by America's largest airline, United, is an excellent example of the daunting scale of the required substitution: the company's contract with a biofuel supplier will provide only 2 percent of the airline's annual fuel consumption. True, today's airliners are increasingly frugal: they are now burning about 50 percent less fuel per passenger-kilometer than they did in 1960. But those savings have been swamped by the continuing expansion of aviation, which has raised annual consumption of jet fuel to more than 250 million tons worldwide.

To meet this demand largely with biojet fuel, we would have to go beyond organic wastes and tap oil-rich seasonal (corn, soybeans, rapeseed) or perennial (palm) oil crops, whose cultivation would require large areas and create environmental problems. Temperate-climate oil crops have relatively low yields: with an average yield of 0.4 tons of biojet per hectare of soybeans, the United States would need to put 125 million hectares—an area bigger than Texas, California, and Pennsylvania combined, or slightly larger than South Africa—under the plow to supply its own jet fuel needs. That's four times the 31 million hectares that the country devoted to soybeans in 2019. Even the highest-yielding option—oil palm, which averages 4 tons of biojet per hectare—would still require more than 60 million hectares of tropical forest to supply the world's aviation fuel. That would necessitate quadrupling the area devoted to palm oil cultivation, leading to the release of carbon accumulated in natural growth.

But why take over huge tracts of land when you can derive biofuels from oil-rich algae? Intensive, large-scale cultivation of algae would require relatively little space and would offer very high productivity. However, Exxon Mobil's experience shows how demanding it would be to scale up to tens of millions of tons of biojet every year. Exxon, working with Craig Venter's Synthetic Genomics, began to pursue this option in 2009, but by 2013, after spending more than $100 million, it concluded that the challenges were too great and decided to refocus on long-term basic research.

As always, the task of energy substitutions would be made easier if we wasted less, say, by flying less. But forecasts are for further substantial growth of air traffic, particularly in Asia. Get used to the unmistakable smell of aviation kerosene—it'll be here for a long time to come, and moreover, it fuels the machines that (as we will see in the next chapter) are extraordinarily safe to fly.

How safe is flying?

You might have thought that 2014 was a bad year for fly-ing. There were four highly publicized accidents: the still-mysterious disappearance of Malaysia Airlines Flight 370 in March; the shooting-down of Malaysia Airlines Flight 17 over Ukraine in July; the Air Algérie Flight 5017 crash in Mali, also in July, with a total of 815 dead; and finally, AirAsia Flight QZ8501 falling into the Java Sea in December.

But according to Ascend, the consulting branch of FlightGlobal that monitors aircraft accidents, 2014 in fact had the best accident rate in history: one fatal acci-dent per 2.38 million flights. True, Ascend did not count the downing of MH Flight 17—which was an act of war, not an accident. Including that incident, as the Inter-national Civil Aviation Organization does in its statistics, raises the rate to 3.0—still much lower than between 2009 and 2011.

And the subsequent years were safer still: fatalities declined to 474 in 2015, 182 in 2016, and to just 99 in 2017. There was a reversal in 2018, with 11 fatal accidents and 514 deaths (still lower than in 2014), including the Lion Air Boeing 737 Max falling into the sea off Jakarta in October. And in 2019, despite another Boeing 737 Max

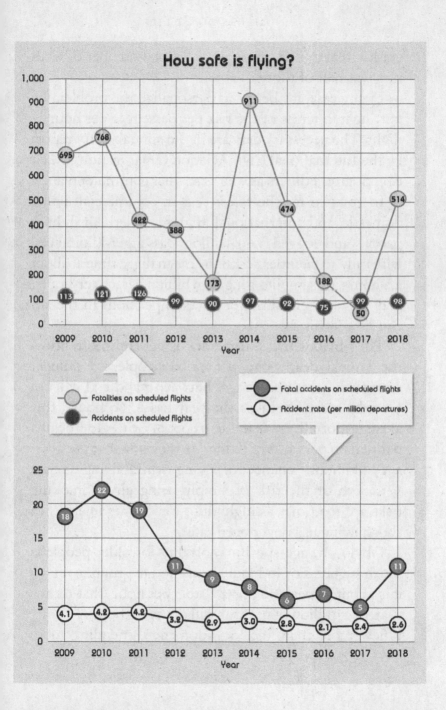

How safe is flying?

Fatalities on scheduled flights
Accidents on scheduled flights
Fatal accidents on scheduled flights
Accident rate (per million departures)

crash—this time in Ethiopia—the total number of fatalities was half of those in 2018.

In any case, it's better to personalize the problem by putting it in terms of the risk per passenger per hour of flight. The necessary data are in the annual safety report by the International Civil Aviation Organization, which covers large jetliners as well as smaller commuter planes.

In 2017, so far the safest year in commercial flying, domestic and international flights carried 4.1 billion people and logged 7.69 trillion passenger-kilometers, with only 50 fatalities. With the mean flight time at about 2.2 hours, this implies roughly 9 billion passenger-hours, and 5.6×10^{-9} fatalities per person per hour in the air. But how low is this risk?

The obvious measuring stick is general mortality—the annual death rate per 1,000 people. In affluent nations that rate now ranges between 7 and 11; I will use 9 as the mean. Because the year has 8,760 hours, this average mortality prorates to 0.000001 or 1×10^{-6} deaths per person per hour of living. This means that the average additional chance of dying while flying is just 5/1,000th of the risk of simply being alive. Smoking risks are 100 times as high; ditto for driving in a car. In short, flying has never been safer.

Obviously, age-specific mortality for older people is much higher. For individuals of my age group (over 75) it is about 35 per 1,000 or 4×10^{-6} per hour (that means that, of a million of us, four will die every hour). In 2017, I flew more than 100,000 kilometers, spending more than 100 hours aloft in large jets belonging to four major airlines whose last fatal accidents were, respectively, in 1983, 1993, 1997, and 2000. In every hour aloft the probability of my demise wasn't even 1 percent higher than it would have been had I stayed on the ground.

Of course, I've had white-knuckle moments. The most recent one was in October 2014, when my Air Canada Boeing 767 headed into the turbulent fringes of a mega-typhoon that was crossing over Japan.

But I never forget that quiet hospital rooms are what should really be avoided. Although the latest assessment of preventable medical errors has greatly reduced the previously exaggerated claims of that risk, hospitalizations remain associated with increased exposure to bacteria and viruses, elevating the risks of hospital-acquired infections, particularly among the elderly. So keep on flying, and avoid hospitals!

Which is more energy efficient—planes, trains, or automobiles?

I have no animosity toward cars and planes. For decades I have depended for local travel on a succession of reliable Honda Civics, and for years I have flown intercontinentally at least 100,000 kilometers annually. At these two extremes—a drive to an Italian food store; a flight from Winnipeg to Tokyo—cars and planes rule.

Energy intensity is the key. When I'm the only passenger in my Civic, it requires about 2 megajoules per passenger-kilometer for city driving. Add another passenger and that figure drops to 1 MJ/pkm, comparable to a half-empty bus. Jet airliners are surprisingly efficient, commonly requiring around 2 MJ/pkm. With full flights and the latest airplane designs, they can do it at less than 1.5 MJ/pkm. Of course, public-transit trains are far superior: at high passenger loads, the best subways need less than 0.1 MJ/pkm. But even in Tokyo, which has a dense network of lines, the nearest station may be more than a kilometer away, too far for many less-mobile people.

But none of these modes of transportation can equal the energy intensity of intercity high-speed trains. These are typically on routes of 150–600 kilometers. Older models of Japan's pioneering bullet train, the *shinkansen* (meaning "new main line"), had an energy intensity of around

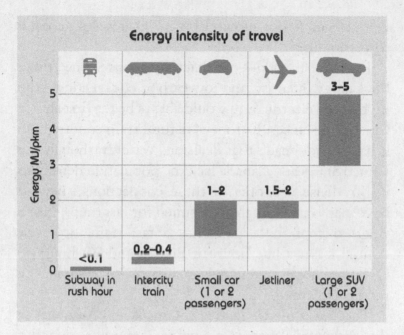

Energy intensity of travel

o.35 MJ/pkm; more recent fast-train designs—the French TGV and German ICE—typically need just o.2 MJ/pkm. That's an order of magnitude less than airplanes.

No less important, high-speed trains are indeed fast. The Lyon–Marseille TGV covers 280 kilometers in 100 minutes, downtown to downtown. In contrast, the scheduled commercial flight time for about the same distance—300 kilometers from New York's LaGuardia Airport to Boston's Logan Airport—is 70 minutes. Then you must add at least another 45 minutes for checking in, 45 minutes for the ride from Manhattan to LaGuardia, and 15 minutes for the ride from Logan to downtown Boston. That raises the total to 175 minutes.

In a rational world—one that valued convenience, time, low energy intensity, and low carbon conversions—the high-speed electric train would always be the first choice for such distances. Europe is natural train country, and it has already made that decision. Yet even though the United States and Canada lack the population density to justify dense networks of these connections, they do have many city pairs that are suited for fast trains. Not a single one of those pairs has a fast train, however. Amtrak's Acela line between Boston and Washington, DC, does not even remotely qualify, as it averages just a measly 110 km/h.

This leaves the US (and also Canada and Australia) as the outstanding laggards in rapid train transportation. But there was a time when America had the best trains in the world. In 1934, 11 years after General Electric made its first diesel locomotive, the Chicago, Burlington and Quincy Railroad began to run its streamlined stainless-steel *Pioneer Zephyr*, a 600-horsepower (447-kilowatt), eight-cylinder, two-stroke diesel-electric unit. This power made it possible for the *Zephyr* to beat the speeds of today's Acela, with an average of 124 km/h on the more than 1,600-kilometer-long run from Denver to Chicago. But there is now no realistic hope that the US could ever catch up with China: at 29,000 kilometers of high-speed rail that country now has the world's longest network of rapid trains, connecting all major cities in its populous eastern half.

FOOD
Energizing Ourselves

The world without synthetic ammonia

By the end of the 19th century, advances in chemistry and plant physiology made it clear that nitrogen is the most important macronutrient (element needed in relatively large amounts) in crop cultivation. Plants also require phosphorus and potassium (the other two macronutrients) and various micronutrients (elements ranging from iron to zinc, all needed in small amounts). A good harvest of Dutch wheat (9 tons per hectare) will contain about 10 percent protein or 140 kilograms of nitrogen, but only about 35 kilograms each of phosphorus and potassium.

Traditional farmers supplied the needed nitrogen in two ways: by recycling any available organic materials (straw, stalks, leaves, human and animal waste) and by rotating grain or oil crops with leguminous plants (cover crops such as alfalfa, clovers, and vetches; and food crops such as soybeans, beans, peas, and lentils). These plants are able to supply their own nitrogen because bacteria attached to their roots can "fix" nitrogen (convert it from the inert molecule in the air to ammonia that is available to growing plants) and they also leave some of it behind for the following grain or oil crop.

The first option was laborious, especially the collection

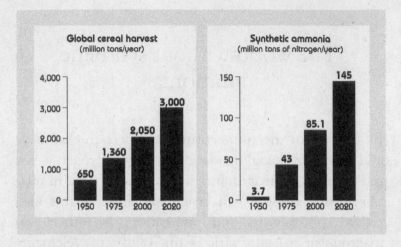

of human and animal waste and its fermentation and application to fields, but manures and night soil had relatively high nitrogen content (commonly 1–2 percent) compared to the less than 0.5 percent nitrogen in straw or plant stalks. The second option required crop rotation and prevented continuous cultivation of staple grain crops, be it rice or wheat. As the demand for staple grains grew with an expanding (and urbanizing) population, it became clear that farmers would not be able to meet future food needs without new, synthetic sources of "fixed" nitrogen—that is, nitrogen available in forms that can be tapped by growing crops.

The quest to do so succeeded by 1909, when Fritz Haber, professor of chemistry at Karlsruhe University, demonstrated how ammonia (NH_3) could be made under high pressure and high temperature in the presence of a metal catalyst. The First World War and the economic crisis of the 1930s slowed down the worldwide adoption of the Haber-Bosch process, but the food needs of the growing global population (from 2.5 billion in 1950 to 7.75 billion in 2020) ensured its massive expansion, from less than 5 million tons in 1950 to about 150 million in recent years. Without this critical input it would have been impossible to multiply staple grain yields (see MULTIPLYING WHEAT YIELDS, p. 226) and feed today's global population.

Synthetic nitrogenous fertilizers derived from Haber-Bosch ammonia (solid urea is the most common product) currently provide roughly half of all nitrogen required by the world's crops, with the rest supplied by rotations with leguminous crops, organic recycling (manures and crop residues), and atmospheric deposition. Because crops now supply about 85 percent of all food protein (with the rest coming from grazing and aquatic foods), this means that without synthetic nitrogen fertilizers, we could not secure enough food for the prevailing diets of just over 3 billion people—more than the combined population of China (where synthetic nitrogen already provides in excess of 60 percent of all inputs) and India. And with the growing populations in parts of Asia and in all of Africa, the shares of humanity dependent on synthetic nitrogen will soon rise to 50 percent.

China still makes some ammonia by using coal as the feedstock, but elsewhere the Haber-Bosch process is based on taking nitrogen from the air and hydrogen from natural gas (mostly CH_4), and also using the gas to supply the high energy requirements of the synthesis. As a result, worldwide synthesis of ammonia and the subsequent production, distribution, and application of solid and liquid nitrogenous fertilizers are now responsible for about 1 percent of global greenhouse gas emissions—and we do not have any commercial non-carbon alternative that could be deployed soon on the required mass scale of making nearly 150 million tons of NH_4 a year.

Immediately more worrisome are large nitrogen losses (volatilization, leaching, and denitrification) resulting from fertilizer use. Nitrates contaminate fresh waters and coastal seas (causing the expansion of dead zones); the atmospheric deposition of nitrates acidifies natural ecosystems; and nitrous oxide (N_2O) is now the third-most important greenhouse gas following CO_2 and CH_4. A recent global assessment concluded that the nitrogen utilization efficiency has actually declined since the early 1960s to around 47 percent—more than half of the applied fertilizer is lost rather than incorporated into harvested crops.

Demand for synthetic nitrogen is saturating in affluent countries, but large increases will be needed to feed some 2 billion people to be born during the next 50 years in Africa. In order to cut future nitrogen losses, we should do everything possible to improve the efficiency of fertilization, reduce food waste (see

THE INEXCUSABLE MAGNITUDE OF GLOBAL FOOD WASTE, p. 230), and adopt moderate meat consumption (see RATIONAL MEAT-EATING, p. 251). And even that will not eliminate all nitrogen losses—but that is the price we pay for having gone from 1.6 billion people in 1900 toward 10 billion by 2100.

Multiplying wheat yields

What is the average wheat yield in central France, in eastern Kansas, or in southern Hebei province? Few people besides the farmers, those who sell them machines and chemicals, the agronomists who counsel them, and the scientists who develop new crop varieties have the answers ready. This is because all but a tiny share of the population in modern societies has become almost completely detached from anything to do with crop cultivation. Except, of course, for eating the products: every crunchy baguette and every croissant, every hamburger bun and every pizza, every steamed bun (*mantou*) and every strand of twisted and stretched *lamian* noodles starts with wheat.

But even those people who consider themselves well educated and widely informed, and who could point out the improved performance of cars or the rising capabilities of computers or cellphones, would have no idea if the average 20th-century staple grain crops had tripled, quintupled, or risen by an order of magnitude. And yet it is these multiples—not those of mobile phone capability or cloud storage—that have made it possible to nearly quintuple the global population between 1900 and 2020 . . . so what happened to the yields of wheat, the world's dominant staple?

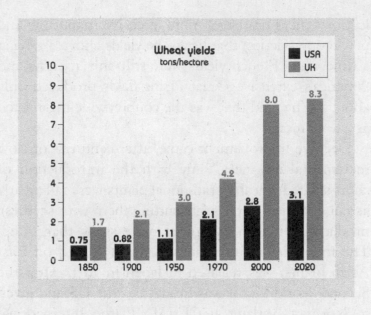

Its traditional yields were both low and highly vari-able, but reconstructions of long-term trends remain arguable. This is the case even with a relatively well-documented (for nearly a millennium) history of English wheat yields, which were usually expressed as returns on planted seed. Following a poor harvest, up to 30 percent of the yield had to be saved for the next year's seed, and the share was commonly no less than 25 percent. Early medieval harvests were often as low as 500–600 kilo-grams per hectare (that is as little as 0.5 ton). Yields up to 1 ton per hectare became common only by the 16th century, and by 1850 the mean was about 1.7 t/ha—roughly a tripling since 1300. Then came a combination of measures (crop rotation including nitrogen-fixing legumes, field drainage, more intensive manuring, and new crop varieties) that lifted the yields above 2 t/ha at a time when French yields were still only 1.3 t/ha and America's extensive Great Plains fields produced only about 1 t/ha (and that was the countrywide mean even as late as 1950!).

Decisive improvement came, after centuries of slow incremental advances, only with the introduction of short-stalked wheats. Traditional plants were tall (nearly as tall as Bruegel's peasants cutting them with scythes), producing three to five times more straw than grain. The first modern short-stalked wheat (based on East Asian plants) was released in Japan in 1935. After the Second World War it was brought to the US and given to Norman Borlaug at CIMMYT (the International Maize and Wheat Improvement Center in Mexico), and his team produced two high-yielding semi-dwarf varieties (yielding as much grain as straw) in 1962. Borlaug won the Nobel Prize; the world got unprecedented harvests.

Between 1965 and 2017, the average global wheat yield almost tripled, from 1.2 to 3.5 t/ha; the Asian average more than tripled (from 1 to 3.3 t/ha), the Chinese mean more than quintupled (from 1 to 5.5 t/ha), and the Dutch mean, already extraordinarily high two generations ago, more than doubled from 4.4 to 9.1 t/ha! In that time, the global wheat harvest nearly tripled to almost 775 million tons, while the population increased 2.3-fold, raising the average per capita supply by about

25 percent and keeping the world comfortably supplied with wheat flour for crusty German *Bauernbrot* (made with wheat and rye flour), Japanese *udon* noodles (wheat flour, a little salt, water) and classic French *mille-feuille* (the puff pastry needed for the sheets is just flour and butter and a bit of water).

But there are concerns. Average wheat yields have been leveling off not only in the EU countries with the highest productivity, but also in China, India, Pakistan, and Egypt, where they remain well below the EU mean. The reasons range from environmentally driven restrictions on the use of nitrogenous fertilizers to water shortages in some regions. At the same time, wheat yields should be benefiting from higher atmospheric CO_2 levels, and agronomic improvements should close some of the yield gaps (differences between a region's yield potential and actual productivity). But, in any case, we would need substantially less wheat if we were to be able—finally—to reduce our indefensibly high food waste.

The inexcusable magnitude of global food waste

The world is wasting food on a scale that must be described as excessive, inexcusable, and, given all of our other concerns about the state of the global environment and quality of human life, outright incomprehensible. The UN's Food and Agricultural Organization puts the annual global losses at 40–50 percent for root crops, fruits, and vegetables, 35 percent for fish, 30 percent for cereals, and 20 percent for oilseeds, meat, and dairy products. This means that, globally, at least one-third of all harvested food is wasted.

Reasons for wasting food differ. In the poorest countries it is most often because of poor storage (as rodents, insects, and fungi feast on improperly stored seeds, vegetables, and fruits) or lack of refrigeration (causing rapid spoilage of meat, fish, and dairy products). That is why in sub-Saharan Africa, most waste takes place even before the food reaches consumers. In the affluent world, however, the main cause is simply the gap between excessive production and actual consumption: despite their high frequency of overeating, most high-income nations provide their citizens with a food supply that would be, on average, adequate for hard-working lumberjacks or coal miners, not for largely sedentary and aging populations.

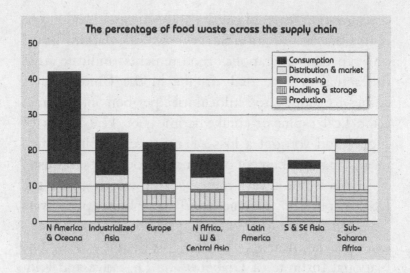

Not surprisingly, the United States is a leading offender, and we have plenty of information to quantify the excess. Average daily food supply in the US comes to about 3,600 kilocalories per person. That's supply, not consumption— and a good thing, too.

Consider that if you omit babies and housebound octogenarians, whose daily requirements are less than 1,500 kilocalories, that would leave more than 4,000 kilocalories available for adults: Americans may eat too much, but they could not all eat that much every day. The United States Department of Agriculture (USDA) adjusts these figures for "spoilage and other waste," and puts the actual daily average available for consumption at about 2,600 kilocalories per person. But even that isn't quite right. Both surveys of self-reported food consumption (carried out by the National Health and Nutrition Examination Survey) and calculations based on expected metabolic requirements indicate that the actual average daily intake in the United States comes to about 2,100 kilocalories per person. Subtract 2,100 kcal/capita of intake from 3,600 kcal/capita of supply and you get a loss of 1,500 kcal/capita, which means about 40 percent of American food goes to waste.

This was not always the case. In the early 1970s, the USDA put the average food availability per capita (adjusted for pre-retail waste) at less than 2,100 kilocalories per day, nearly 25 percent less than it is now. The National Institute of Diabetes and Digestive and Kidney Diseases estimates that the United States' per capita food waste increased by 50 percent between 1974 and 2005, and that the problem has gotten worse since then.

But even if the average US daily loss had remained at 1,500 kilocalories per capita, a simple calculation shows that in 2020 (with about 333 million people) this wasted food could have provided adequate nutrition (2,200 kilocalories per capita) to about 230 million people, slightly more than the entire population of Brazil, Latin America's largest nation and the world's sixth-most populous country.

Yet even as they waste food, Americans are still eating far more of it than is good for them. The prevalence of obesity—defined as a body mass index of 30 or greater—more than doubled between 1962 and 2010, rising from 13.4 percent to 35.7 percent among adults over age 20. Add to this number the merely overweight (a BMI between 25 and 30) and you find that, among adults, 74 percent of males and 64 percent of females have an excessively high weight. Most worrisome, as obesity is usually a lifelong condition, that proportion is now above 50 percent for children above the age of six as well.

The UK's Waste and Resources Action Programme (WRAP) provides different perspectives by tracing the phenomenon in unusual detail. In Britain, total food waste amounts to about 10 million tons a year and it is worth about £15 billion (or nearly $20 billion), but inedible parts (skins, peelings, bones) make up only 30 percent of that total—so 70 percent of wasted food could have been eaten! WRAP also documented the whys of the process: nearly 30 percent of the waste is due to "not being used in time," a third because of the expiry of "best before" dates, about 15 percent because too much was cooked or served, and the rest is due to other reasons, including personal preferences, fussy eating, and accidents.

Food loss goes beyond the wasted nutrition, however—it inevitably entails a significant waste of labor and energy used directly to operate field machinery and irrigation pumps, and indirectly to produce the steel, aluminum, and plastics needed to make those mechanical inputs and to synthesize fertilizers and pesticides. The extra agricultural effort also ends up hurting the environment by causing soil erosion, nitrate leaching, the loss of biodiversity, and the growth of antibiotic-resistant bacteria; and the production of wasted food may be responsible for as much as 10 percent of global greenhouse gas emissions.

Affluent countries need to produce considerably less food and consume it with considerably less waste. And yet the mantra of higher food production is chanted as loudly as ever. Its most recent permutation is to produce more by eventually flooding the markets with fake meat made from altered legume proteins. Instead, why not try to find clever ways to reduce food waste to a more acceptable level of loss? Cutting food waste in half would lead the way to a more rational use of food worldwide, and the benefits could be huge: WRAP estimates that a dollar invested in food waste prevention has a 14-fold return in associated benefits. Is this not persuasive enough?

The slow *addio* to the Mediterranean diet

The benefits of the Mediterranean diet became widely known after 1970, when Ancel Keys published the first installment in his long-term study of nutrition and health in Italy, Greece, and five other countries, and found it was associated with a low incidence of heart disease.

The key traits of the diet are a high intake of carbohydrates (mostly bread, pasta, and rice) complemented by pulses (beans, peas, chickpeas) and nuts, dairy products (mostly cheese and yogurt), fruits and vegetables, seafood, and lightly processed seasonal foods, generally cooked with olive oil. It also includes much more modest quantities of sugar and meat. Best of all, plenty of wine is taken with the food. The latter habit is not one that is now recommended by dieticians, but it is clear that the Mediterranean diet reduces the risk of cardiovascular problems, cuts the risk of certain cancers by about 10 percent, and offers some protection against type 2 diabetes. There is little doubt that if Western countries had followed it en masse, they'd never have reached the levels of obesity prevalent today. In 2013, UNESCO inscribed the diet on the list of the Intangible Cultural Heritage of Humanity, with Croatia,

Cyprus, Greece, Italy, Morocco, Portugal, and Spain as the designated countries.

Yet even in those havens of health there is a growing problem: the true Mediterranean diet is now eaten only in certain isolated coastal or mountainous outposts. The dietary transition has been rapid and far-reaching, particularly in the two most populous countries of the region, Italy and Spain.

During the past 50 years, the Italian diet has become *more* Mediterranean only for fruit, the consumption of which went up by nearly 50 percent. Meanwhile, consumption of animal fats and meat trebled. Olive oil now supplies less than half of all dietary fats and—*incredibile!*—the consumption of pasta is down and that of wine is way down, in a drop of about 75 percent. Italians now buy as much beer as they do *rosso* and *bianco*.

The Spanish retreat from the Mediterranean diet has been even faster and even more complete. Spaniards still like their seafood, the consumption of which has increased, but they've moved away from grains, vegetables, and legumes. Olive oil now provides less than half of all the country's fats. And, remarkably, Spaniards on average now drink only around 20 liters of wine per year—which is less than half the beer they consume. That's comparable to what you see in Germany and the Netherlands!

Could there be a more potent symbol of the diet's demise than *tinto* beaten by *cerveza*? And even most Europeans (keeping old dietary patterns in their memory) are unaware that the Spanish per capita meat supply, only at about 20 kilograms a year when Franco died in 1975, is now, at nearly 100 kilograms, well ahead of such traditionally carnivorous nations as Germany, France, and Denmark.

And the prospects are not good. A new pattern of eating has become the norm among the young people, who also buy less fresh food than their parents did. Spain, for example, has no shortage of McDonald's, KFCs, Taco Bells, and Dunkin' Donuts—or Dunkin' Coffee, as it's called there. The global reach of meaty, fatty, salty, and sugary fast food is doing away not just with an ancient culinary heritage, but also with one of the few advantages the ancient world had over the modern one.

Reasons for this shift have been universal. Higher incomes allow higher meat, fat, and sugar intakes. Traditional families have been replaced by two-income and single-person households that cook less at home and buy more ready-to-eat meals. And busier lifestyles promote snacking and convenience food. It's no wonder that obesity rates have been increasing in Spain and Italy, as well as in France.

Bluefin tuna: On the way to extinction

Consider the tuna ... Its near-perfect hydrodynamics and efficient propulsion, powered by warm-blooded muscles deep within the body, make it an outstanding swimmer. The largest ones top 70 kilometers per hour, or around 40 knots—fast for a powerboat, and far faster than any known submarine.

But their size and tasty meatiness have put the most majestic of these fish on the road to extinction. The white meat you get in cans comes from the relatively abundant albacore—a small fish, typically less than 40 kilograms (red canned meat comes from the abundant skipjack, another small tuna). In contrast, bluefin (in Japanese, *maguro* or *hon-maguro*, "true tuna") has always been the rarest tuna. Adults can grow to more than 3 meters and weigh more than 600 kilograms.

The bluefin is Japan's first choice for sashimi and sushi. When these dishes became popular in Edo (Tokyo) during the 19th century, the choice cuts originally came from the less oily, red inner muscles (*akami*); later the preference shifted to cuts from body sides below the midline (fatty *chūtoro*) and from the fish's belly (extra-fatty *ōtoro*). Exceptional bluefins have been sold for exceptional prices at Tokyo's New Year's auctions.

Another record price for a bluefin tuna

The latest record was set in 2019: $3.1 million for a 278-kilogram fish caught off northern Japan. That's more than $11,100/kg!

Japan consumes about 80 percent of the worldwide bluefin catch, far more than its own allowed quota, and to fill the gap bluefins are now imported to Japan either fresh, as air cargo, or gilled, gutted, and frozen solid. Rising demand is increasingly satisfied by fish caught in the wild and then fattened in cages, where they're fed sardines, mackerel, and herring. The demand is reaching new highs as the sushi craze has turned a Japanese favorite into a global status food.

The reported worldwide catch of three bluefin species is now about 75,000 tons a year. That's less than it was 20 or 40 years ago, but illegal catches and underreported landings, widespread and constant for decades, remain substantial. A pioneering comparison of logbooks of Japan's tuna-fishing fleet (thought to be highly accurate) and tuna sold in Japan's largest fish markets showed at least a twofold discrepancy.

The principal fishing nations have resisted any deep cuts in their fishing quotas. Therefore, the only way to ensure long-term survival is to stop the trade in the most endangered stocks. In 2010, the World Wildlife Fund, fishery experts at the UN's Food and Agriculture Organization, and the Principality of Monaco asked for an international trade ban on the northern bluefin, but the proposal was defeated. Moreover, it might be too late for even a total fishing ban in the Mediterranean and in the northeast Atlantic to prevent the collapse of those bluefin fisheries.

And, unfortunately, it's very hard to raise bluefins from eggs on a sea ranch, as it were, because most of the tiny, fragile larvae do not survive the first three or four weeks of life. The most successful Japanese operation, Kindai University's Fisheries Laboratory, has worked for some 30 years to master the process, but even so only 1 percent of the fish survive to maturity.

Declining catches and farming challenges have resulted in rampant mislabeling around the world, and particularly in the United States. There is a very high chance that you are eating another species rather than any tuna that's listed on your restaurant's menu: In the US more than half of all tuna served in restaurants and sushi shops is mislabeled!

Why chicken rules

For generations, beef was the United States' dominant meat, followed by pork. When annual beef consumption peaked in 1976 at about 40 kilograms (boneless weight) per capita, it accounted for nearly half of all meat; chicken had just a 20 percent share. But chicken caught up by 2010, and in 2018 chicken's share came to 36 percent of the total, nearly 20 percentage points higher than beef. The average American now eats 30 kilograms of boneless chicken every year, bought overwhelmingly as cut-up or processed parts (from boneless breasts to Chicken McNuggets).

The constant obsession with diet in the United States—in this case the fear of dietary cholesterol and saturated fat in red meat—has been a factor in the shift. The differences, however, are not striking: 100 grams of lean beef has 1.5 grams of saturated fat, compared with 1 gram in skinless chicken breast (which actually has more cholesterol). But the main reason for chicken's ascendance has been its lower price, which reflects its metabolic advantage: no other domesticated land animal can convert feed to meat as efficiently as broilers, as chickens bred and raised specifically for meat production are known.

Why chicken rules

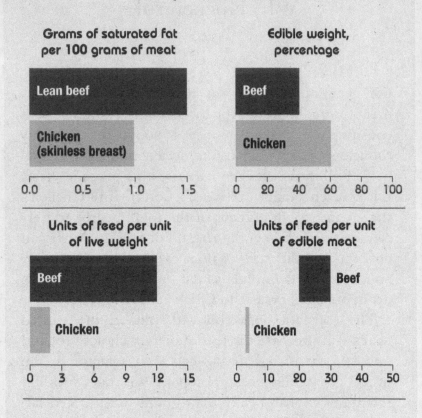

Grams of saturated fat per 100 grams of meat

Lean beef

Chicken (skinless breast)

0.0   0.5   1.0   1.5

Edible weight, percentage

Beef

Chicken

0   20   40   60   80   100

Units of feed per unit of live weight

Beef

Chicken

0   3   6   9   12   15

Units of feed per unit of edible meat

Beef

Chicken

0   10   20   30   40   50

Average feed-to-meat conversion efficiencies

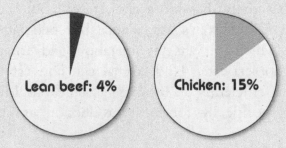

Lean beef: 4%

Chicken: 15%

Modern breeding advances have a lot to do with this efficiency.

During the 1930s, the average feeding efficiency for broilers (at about 5 units of feed per unit of live weight) was no better than for pigs. That feeding rate was halved by the mid-1980s, and the latest US Department of Agriculture's feed-to-meat ratios show that it now takes only about 1.7 units of feed (standardized in terms of feed corn) to produce a unit of broiler live weight (before slaughter), compared with nearly 5 units of feed for hogs and almost 12 units for cattle.

Because edible weight as the share of live weight differs substantially among the leading meat species (it is about 60 percent for chicken, 53 percent for pork, and only about 40 percent for beef), recalculations in terms of feeding efficiencies per unit of edible meat are even more revealing. Recent ratios have been 3–4 units of feed per unit of edible meat for broilers, 9–10 for pork, and 20–30 for beef. These numbers correspond to average feed-to-meat conversion efficiencies of, respectively, 15, 10, and 4 percent.

In addition, broilers have been bred to mature faster and to put on an unprecedented amount of meat. Traditional free-running birds were slaughtered at the age of one, when they weighed only about 1 kilogram. The average weight of American broilers rose from 1.1 kilograms in 1925 to nearly 2.7 in 2018, while the typical feeding span was cut from 112 days in 1925 to just 47 days in 2018.

Consumers benefit while the birds suffer. They gain

weight so rapidly because they can eat as much as they want while being kept in darkness and strict confinement. Because consumers prefer lean breast meat, the selection for excessively large breasts shifts the bird's center of gravity forward, impairs its natural movement, and puts stress on its legs and heart. But the bird cannot move anyway; according to the National Chicken Council, a broiler is allotted just 560–650 square centimeters, an area only slightly larger than a sheet of standard A4 letter paper. As long periods of darkness improve growth, broilers mature under light intensities resembling twilight. This condition disrupts their normal circadian and behavioral rhythms.

On one side, you have shortened lives (less than seven weeks for a bird whose normal lifespan is up to eight years) and malformed bodies in dark confinement; on the other, in late 2019 you got retail prices of about $2.94 per pound ($6.47/kg) for boneless breast, compared with $4.98/lb for round beef roast and $8.22/lb for choice sirloin steak.

But chicken's rule hasn't yet gone global; thanks to its dominance in China and in Europe, pork is still about 10 percent ahead worldwide, while beef is the leading meat in most countries in South America. Still, broilers mass-produced in confinement will, almost certainly, come out on top worldwide within a decade or two. Given this reality, consumers should be willing to pay a bit more in order for the producers to make the short lives of broilers less stressful.

(Not) drinking wine

France and wine, what an iconic link—and, for centuries, how immutable! Introduced by Greeks long before the Romans conquered Gallia, greatly expanded during the Middle Ages, and eventually becoming a symbol of quality (Bordeaux, Bourgogne, Champagne) at home and abroad, French viticulture, wine-drinking, and wine exports have been long established as one of the key signifiers of national identity. The country always produced and drank copiously, with farmers and villagers in wine regions consuming their own vintages, and towns and cities enjoying a wide selection of tastes and prices.

Regular statistics of annual French per capita consumption of wine begin in 1850 with a high mean of 121 liters a year—nearly two medium glasses (175 milliliters) a day. By 1890, phylloxera infestation (it began in 1863) had cut the country's grape harvest by nearly 70 percent compared to the peak in 1875, and French vineyards had to be reconstituted by grafting on resistant (mostly American) rootstocks. As a result, annual consumption of wine fluctuated, but rising imports (in 1887 they were as high as half of the domestic output) prevented any steep decline in the total supply, and eventual vineyard recovery brought the pre–First World

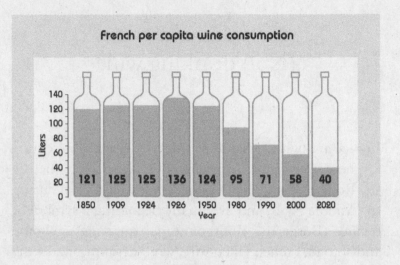

French per capita wine consumption

War per capita peak to 125 liters a year in 1909. That rate was equaled in 1924 and surpassed during the next two years, setting the all-time per capita record of 136 liters a year in 1926, and by 1950 the rate was only slightly lower, at about 124 liters.

Postwar French living standards remained surprisingly low: according to the 1954 census only 25 percent of homes had an indoor toilet, and only 10 percent had a bathtub, shower, or central heating. But all of that changed rapidly during the 1960s, and the rising affluence also brought some notable dietary shifts—and the decline of wine-drinking. By 1980 the per capita annual mean was down to about 95 liters a year, by 1990 it sank to 71 liters, and by the year 2000 it had fallen to just 58 liters, cut by half over the course of the 20th century. The current century has seen further declines, and the latest available data show the mean at just 40 liters a year, 70 per cent below the 1926 record. The wine consumption survey of 2015 (to be repeated in 2020) details deep gender and generational divides that explain the falling trend.

Forty years ago, more than half of French adults were drinking wine nearly every day, but the share of all adults who drink wine regularly is now just 16 percent. More specifically, the share is 23 percent among men and 11 percent among women, and only 1 percent among those aged 15–24 years and 5 percent for 25- to 34-year-olds, compared to 38 percent for people above 65 years of age. Obviously, this gender and generational divide does not hold any promise of rising consumption in the future, and it applies to all alcoholic drinks: beer, liquors and cider have also seen gradual consumption declines, while the beverages with highest average per capita gains include mineral and spring waters (roughly doubling since 1990), fruit juices, and carbonated soft drinks.

As wine-drinking changed from a regular habit to an occasional indulgence, France also lost its historical wine-consumption primacy to Slovenia and Croatia (both at nearly 45 liters a year per capita). But while no other traditional wine-drinking country has seen greater declines—in both absolute and relative terms—than France, Italy has come close, and wine consumption has also decreased in Spain and Greece.

There has been one positive trend, however, as France's wine exports remain strong, with a new record (at about $11 billion) set in 2018. Premium prices commanded by French products are attested to by the fact that they account for 15 percent of the global trade in wines and spirits but 30 percent of the total value. Americans (whose average per capita consumption of wine has been up by more than 50 percent during the past 20 years) have been the largest importers of French wines, and the demand by newly rich Chinese has claimed a growing share of sales.

But in the country that gave the world countless *vins ordinaires* as well as exorbitantly priced *Grands Crus Classés*, the sound of clinking stemmed glasses and wishes of *santé* have become an endangered habit.

Rational meat-eating

Eating meat in general (and beef in particular) has now joined a list of highly undesirable habits, as the long-extant concerns about meat's drawbacks—ranging from supposedly deleterious health effects to extraordinarily high land use and the large water footprint required to grow animal feed—have been joined by near-apocalyptic warnings about methane from cattle as a key driver of global climate change. Realities are far less inflammatory. We are—much like chimpanzees, our closest primate ancestors whose males are keen hunters of smaller animals such as monkeys and young wild pigs—an omnivorous species, and meat has always been an important part of our normal diet. Meat (together with milk and eggs) is an excellent source of complete dietary protein required for growth; it contains important vitamins (above all, those of B-complex) and minerals (iron, zinc, magnesium); and it is a satisfying source of dietary lipids (fats providing the feeling of satiety, and hence highly prized by all traditional societies).

Inevitably, animals, particularly cattle, are inefficient converters of feed into meat (see WHY CHICKEN RULES, p. 243), and affluent countries have expanded their meat production to such an extent that the principal task

The Fat Kitchen: Pieter van der Heyden after Pieter Bruegel

of agriculture has become not to grow crops for people but feed for animals. In North America and Europe, about 60 percent of the total crop harvest is now destined for feeding—not directly for food. This, of course, has major environmental consequences, particularly due to the need for nitrogen fertilizers and water. At the same time, citing the large volumes of water needed to produce feed for cattle is quite misleading. The minimum water requirement per kilogram of boneless beef is, indeed, high, on the order of 15,000 liters, but only about half a liter of that ends up incorporated in the meat, with more than 99 percent being water needed for the growth of feed crops which eventually re-enters the atmosphere via evaporation and plant transpiration, and rains down.

As for the health effects of eating meat, large-scale studies show that moderate meat consumption is not associated with any adverse outcomes—but if you do not trust their methodology you can simply compare national life expectancies (see the next chapter) with average per capita meat consumption. At the top of the longevity list are the Japanese (moderate meat consumers; in 2018, almost exactly 40 kilograms in carcass weight per capita) followed by the Swiss (substantial meat eaters, with more than 70 kilograms), Spaniards (Europe's highest meat consumers, with more than 90 kilograms), Italians (not far behind, with more than 80 kilograms) and Australians (more than 90 kilograms, of which about 20 kilograms is beef). So much for meat and a lack of longevity.

At the same time, the Japanese diet (in fact, the East Asian diet in general) shows there is no additional health or longevity benefit from the high consumption of meat, and that is why I strongly advocate rational meat consumption based on moderate intakes of meat produced with greatly reduced environmental impact. The key components of this global adoption would be to adjust the shares of the three dominant meats. For pork, chicken, and beef respectively they were 40, 37, and 23 percent of the global output of about 300 million tons in 2018; by shifting the split to 40, 50, and 10 percent, we could (thanks to grain feed saved by reducing inefficient beef production) produce easily 30 percent more chicken meat and 20 percent more pork, while more than halving beef's environmental burden—and still supplying at least 10 percent more meat.

The new meat total would be close to 350 million tons, and it would prorate to about 45 kilograms of carcass weight or 25–30 kilograms of edible meat (no bones) for every one of 7.75 billion people inhabiting the planet in early 2020!

This is similar to what an average Japanese person has been recently consuming, but also how much a significant share of the population in France—that meat-eating nation par excellence—now prefers to eat: a recent French study showed that nearly 30 percent of French adults have become *petits consommateurs*, with intakes (edible meat) averaging just 80 grams per day or about 29 kilograms a year.

In nutritional terms, the annual intake of 25–30 kilograms of edible meat would (assuming a 25 percent protein content) supply close to 20 grams of complete protein per day: 20 percent more than the recent mean, yet produced with greatly reduced environmental impact and providing all the health and longevity benefits of moderate carnivory.

So why not follow both the longest-living population and a French smart set in their habits? As in so many other matters, moderation could go a long way . . .

The Japanese diet

Modern Japan: rich on paper, but with cramped housing, long and crowded commutes, working hours stretching into the evenings, short holidays, still too many people smoking, and enormous pressure to conform in a traditionally hierarchical society. There is also the ever-present risk of major earthquakes and (in large parts of the country) volcanic eruptions, and the seasonal threat of massive typhoons and heat waves (not to mention living next to North Korea . . .). And yet, Japanese life expectancy at birth is higher than in any other nation. The latest numbers (females/males, for 2015–20, in years) are 87.5/81.3 for Japan, 86.1/80.6 for Spain, 85.4/79.4 for France, 82.9/79.4 for the UK, and 81.3/76.3 for the US. Even more remarkably, at age 80 a Japanese woman can now expect to live an additional 12 years, compared to 10 years in the US and 9.6 years in the UK.

Could unique genetics explain this? Most unlikely, as the islands had to be settled by migrants from the neighboring continent—and a recent study of fine-scale genetic structure and the evolution of the Japanese population confirms that the expected components of the ancestry profile come above all from the Korean and also from Han Chinese and Southeast Asian clusters.

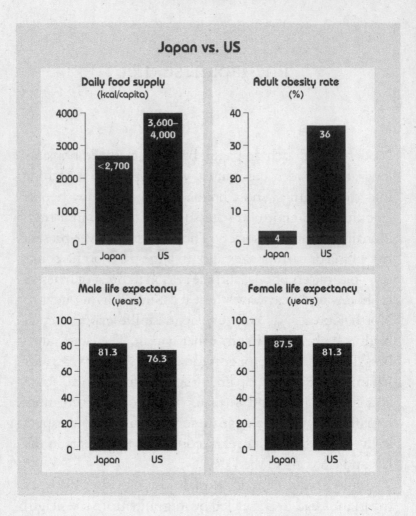

Maybe it is down to widespread and intense religious convictions—to mind over matter? But spirituality rather than religiosity might better describe the Japanese mind-set, and there are no indications that such traditional beliefs are held more deeply there than in other populous nations with old cultural heritage.

Diet, then, should be the best explanation, but which part of it? Focusing on well-known national favorites is hardly helpful. Soy sauce (*shōyu*) is shared with a large part of continental Asia, from Myanmar to the Philippines, as is bean curd (*tōfu*) and, to a lesser extent, even *nattō* (another soybean-based, but fermented, foodstuff). Shades of color may differ, but Japanese green tea— *ryokuchā* or simply *ochā*, the less processed leaves of *Camellia sinensis*—came from China, which still produces and drinks most of it (although less in per capita terms). But food balance sheets (accounts of supply available at retail level and excluding food waste) show important differences in the macronutrient composition of the average Japanese, French, and American diets. Foods of animal origin supply 35 percent of all dietary energy in France and 27 percent in the US, but only 20 percent in Japan.

But this tilt toward a significantly more plant-based diet is less important than the share of food energy coming from fats (lipids, be they of plant or animal origin) and from sugar and other sweeteners. In both the US and France, dietary fat provides almost two times (1.8 to be exact) more food energy than in Japan, while Americans have at their daily disposal nearly 2.5 times more sugar and sweeteners (dominated in the US by high-fructose corn syrup) than the Japanese, with the multiple about 1.5 for France. Always keeping in mind that these are just broad statistical associations, not causal claims, we might conclude that through elimination of likely nutritional factors, we see lower fat and lower sugar intakes as possibly important co-determinants of longevity.

But these two relatively low intakes are a part of what I see as by far the most important explanatory factor, as Japan's true exceptionalism: the country's remarkably moderate average per capita food supply. While food balance sheets of virtually all affluent Western nations (be it the US or Spain, France or Germany) show a daily availability of 3,400–4,000 kilocalories per capita, the Japanese rate is now below 2,700 kilocalories, roughly 25 percent lower. Of course, actual average consumption cannot be at a 3,500-kilocalorie-per-day level (only hard-working, big-stature men might need that much), but even after an indefensibly high share of food waste, this high supply translates into excessive eating (and obesity).

In contrast, studies of actual food intakes show that the Japanese daily mean is now below 1,900 kilocalories, commensurate with age distribution and physical activity of the aging Japanese population. This means that perhaps the single most important explanation of Japan's longevity primacy is quite simple: moderate overall food consumption, the habit expressed in just four kanji characters, 腹八分目 (*hara hachi bun me*, "belly eight parts [in ten] full")—an ancient Confucian precept, and hence yet another import from China. But the Japanese, unlike the banqueting and food-wasting Chinese, actually do practice it.

Dairy products—the counter-trends

Nearly all newborns produce enough lactase, the enzyme needed to digest lactose—the sugar (a disaccharide composed of glucose and galactose) in their mothers' milk. Only a tiny share of babies have congenital lactase deficiency (that is, lactose intolerance). But after infancy, the ability to digest milk diverges. In societies that were originally pastoral or kept domesticated dairy animals, the capacity to digest lactose persists; while in those societies that never kept milking animals, it weakens and even disappears. Typically, this loss translates only into abdominal discomfort after drinking a small amount of milk, but it can cause nausea and even vomiting.

Evolution has produced complex patterns of these traits, with lactase-deficient populations surrounded by milk drinkers (such as the horse milk-drinking Mongolians and yak milk-drinking Tibetans north and west of the non-milk-drinking Chinese), or even with the two societies intermingled (cattle pastoralists and slash-and-burn farmers or hunters of sub-Saharan Africa).

Given these realities, it is remarkable that economic modernization has produced two counterintuitive outcomes: dairy strongholds have seen prolonged declines

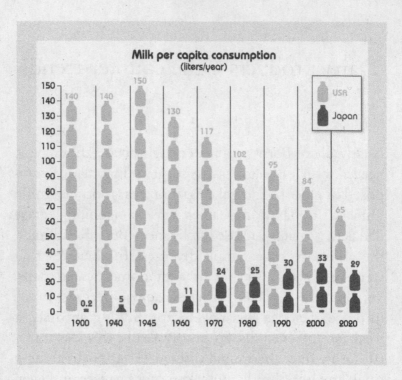

Milk per capita consumption
(liters/year)

of average per capita milk consumption; while in several traditionally non-milk-drinking societies, demand for liquid milk and dairy products has risen from nothing to appreciable quantities. At the beginning of the 20th century, annual US consumption of fresh milk (including cream) was almost 140 liters per capita (80 per cent of it as whole milk); it peaked at about 150 liters in 1945, but the subsequent decline cut it by more than 55 percent, to about 66 liters by 2018. The concurrent decrease of demand for all dairy products has been slower, in large part because of the still slowly rising consumption of mozzarella via American pizza.

The key factors behind the decline have included higher consumption of meat and fish (supplying protein and fat formerly derived from milk) and decades of warnings about the deleterious effect of consuming saturated dairy fat. That conclusion has been disproved, and the latest findings claim that dairy fat may actually lower the frequency of coronary heart disease and stroke mortality—but these findings come too late for the declining industry. A similar retreat took place among Europe's leading dairy consumers, where traditionally high levels of milk drinking were accompanied by daily eating of cheeses. Most notably, the French annual per capita milk consumption was about 100 liters in the mid-1950s, but by 2018 the rate was down to 45 liters.

Japan offers the best example of dairy's rise in a non-milk-drinking society. Annual per capita supply averaged less than 1 liter in 1906, and 5.4 liters by 1941. The latter total prorated to 15 milliliters (or a tablespoon) a day: in reality, this meant that by the time American forces occupied the country in 1945, none but a few large-city dwellers ever drank milk or ate yogurt and cheese. Milk was introduced through the National School Lunch Program in order to eliminate urban/rural discrepancy in childhood growth, and the per capita rates rose to 25 liters per year in 1980 and 33 liters per year by the year 2000, when the total dairy consumption (including cheeses and yogurt) was equivalent to more than 80 liters per year!

Given the country's size, Chinese dairy adoption was necessarily slower, but the average rates rose from negligible minima during the 1950s to 3 liters annually per capita during the 1970s (before the start of China's rapid modernization), and are now more than 30 liters—higher than in South Korea, another traditionally non-milk-drinking culture that now consumes milk, cheeses, and yogurt. Diversification of diets, convenience of dairy foods in modern urban societies, smaller family size, and higher shares of working women in cities have been the main driving factors of this Chinese transition, which was supported by the government's elevation of milk to the status of healthy, prestige food, though it has been marred by poor quality and even outright adulteration: in 2008, some 300,000 babies and children were affected by drinking milk dosed with melamine, an industrial chemical added in order to increase milk's nitrogen and hence its apparent protein content.

But how have lactase-deficient societies been able to undergo this shift? Because lactose intolerance is not universal, and because it is relative rather than absolute. Four-fifths of Japanese have no problems drinking up to a cup of milk a day, and that would translate to annual consumption of more than 70 liters—more than the recent American mean!

And fermentation removes progressively more lactose, with fresh cheeses (such as ricotta) retaining less than a third of the lactose present in milk, and hard varieties (such as Cheddar or Parmigiano) having just a trace. And while yogurt retains nearly all of the original lactose, its bacterial enzymes facilitate the digestion. Milk, an ideal food for babies, is thus also, in moderation, an excellent food for anybody . . . except for those with overt lactose intolerance.

ENVIRONMENT

Damaging and Protecting Our World

Animals vs. artifacts—which are more diverse?

Our count of living species remains incomplete. In the 250-plus years since Carl Linnaeus set up the modern taxonomic system, we have classified around 1.25 million species, about three-quarters of them animals. Another 17 percent are plants, and the remainder are fungi and microbes. And that's the official count—the number of still-unrecognized species could be several times higher.

The diversity of man-made objects is easily as rich. Although my comparisons involve not just those proverbial apples and oranges but apples and automobiles, they still reveal what we have wrought.

I will construct my taxonomy of all man-made objects by creating a classification analogical to that of living organisms. The domain of all human designs is equivalent to the Eukarya (all living organisms that have nuclei in their cells), which contains the three large kingdoms of fungi, plants, and animals. I posit that the domain of all man-made objects contains a kingdom of complex, multicomponent designs, equivalent to the kingdom of animals. Within that kingdom we have the phylum of designs powered by electricity, equivalent to the Chordata, creatures with a dorsal nerve cord. Within that

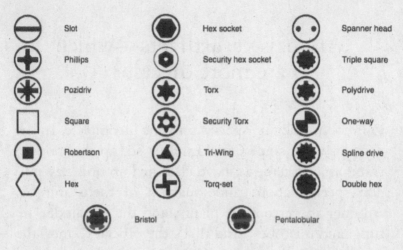

Screwdriver heads: everyday example of diversity in design

phylum is a major class of portable designs, equivalent to mammals. Within that class is the order of communications artifacts, equivalent to Cetacea, the class of whales, dolphins, and porpoises, and it contains the family of phones, equivalent to Delphinidae, the oceanic dolphins.

Families contain genera, such as *Delphinus* (common dolphin), *Orcinus* (orcas), and *Tursiops* (bottlenose dolphins). And, according to GSM Arena, which monitors the mobile phone industry, in early 2019 there were more than 110 cellphone genera (brands). Some genera contain a single specific species—for instance, *Orcinus* contains only *Orcinus orca*, the killer whale. Other genera are species-rich. In the cellphone realm, none is richer than Samsung, which now includes nearly 1,200 devices. It is followed by LG, with more than 600, and Motorola and Nokia, each with almost 500 designs. Altogether, in early 2019 there were some 9,500 different mobile "species"—and that total is considerably larger than the known diversity of mammals (fewer than 5,500 species).

Even if we were to concede that cellphones are just varieties of a single species (like the Bengal, Siberian, and Sumatran tigers), there are many other numbers that illustrate how species-rich our designs are. The World Steel Association lists about 3,500 grades of steel, more than all known species of rodents. Screws are another super-category: add up the combinations based on screw materials (aluminum to titanium), screw types (from cap to drywall, from machine to sheet metal), screw heads (from washer-faced to countersunk), screw drives (from slot to hex, from Phillips to Robertson), screw shanks and tips (from die point to cone point), and screw dimensions (in metric and other units), and you end up with many millions of possible screw "species."

Taking a different tack, we have also surpassed nature in the range of mass. The smallest land mammal, the Etruscan shrew, weighs as little as 1.3 grams, whereas the largest, the African elephant, averages about 5 tons. That's a range of six orders of magnitude. Mass-produced cellphone vibrator motors match the shrew's weight, while the largest centrifugal compressors driven by electric motors weigh around 50 tons, for a range of seven orders of magnitude.

The smallest bird, the bee hummingbird, weighs about 2 grams, whereas the largest flying bird, the Andean condor, can reach 15 kilograms, for a range of nearly four orders of magnitude. Today's miniature drones weigh as little as 5 grams, versus a fully loaded Airbus 380 which weighs 570 tons—a difference of eight orders of magnitude.

And our designs have a key functional advantage: they can work and survive pretty much on their own, unlike our bodies (and those of all animals), which depend on a well-functioning microbiome—there are at least as many bacterial cells in our guts as there are cells in our organs. That's life for you.

Planet of the cows

For years I have tried to imagine how Earth would appear to a comprehensive and discerning probe dispatched by wonderfully sapient extraterrestrials. Of course, the probe would immediately conclude, after counting all organisms, that most individuals are either microscopic (bacteria, archaea, protists, fungi, algae) or very small (insects), but also that their aggregate weight dominates the planetary biomass.

That would not really be surprising. What these tiny creatures lack in size, they more than make up in numbers. Microbes occupy every conceivable niche of the biosphere, including many extreme environments. Bacteria account for about 90 percent of the human body's living cells, and as much as 3 percent of its total weight. What would be surprising, however, is the picture the probe would paint of the macroscopic forms of animal life, which is dominated by just two vertebrates—cattle (*Bos taurus*) and humans (*Homo sapiens*), in that order.

Unlike the extraterrestrial scientists, we do not get an instant readout. Even so, we can quantify cattle zoomass and human biomass (anthropomass) with a fair degree of accuracy. The numbers of large, domesticated ruminants in all high-income countries are known, and they

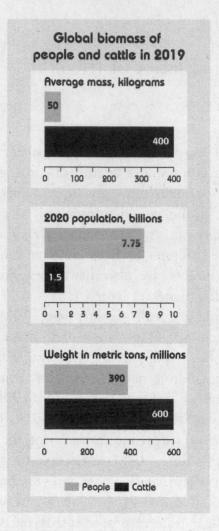

can be reasonably estimated for all low-income and even pastoral societies. The UN's Food and Agricultural Organization puts the global cattle count at about 1.5 billion head in 2020.

To convert these numbers into living ruminant zoomass requires adjustments for age and sex distribution. Large bulls weigh more than 1,000 kilograms; American beef cows are slaughtered when they reach nearly 600 kilograms, but Brazilian cattle go to market at less than 230 kilograms; and India's famous Gir milk breed weighs less than 350 kilograms when mature. A good approximation is to assume an average sex- and age-weighted body mass of 400 kilograms; that implies a total live cattle zoomass of some 600 million tons.

Similarly, when calculating the total mass of humanity it is necessary to consider the age and body weights of populations. Low-income countries have much higher shares of children than affluent nations (in 2020, about 40 percent in Africa compared to about 15 percent in Europe). At the same time, the rates of overweight and obese people range from the negligible (in Africa) to 70 percent of the adult population (United States). That is why I am using specific means for different continents, derived from available population age and sex structures as well as from anthropometric studies and growth curves for representative countries. That complex adjustment produces a weighted mean of about 50 kilograms per capita—which, given a total of 7.75 billion people, implies a global anthropomass of nearly 390 million tons in 2020.

This means that the cattle zoomass is now more than 50 percent larger than the anthropomass, and that the live weight of the two species together is very close to a billion tons. Even the largest wild mammals add up to only a small fraction of those masses: the 350,000 elephants in Africa, with an average body weight of 2,800 kilograms, have an aggregate zoomass of less than 1 million tons, which is less than 0.2 percent of the cattle zoomass. By 2050 there will be 9 billion people and, most likely, 2 billion cattle, together augmenting their already crushing dominance of Earth.

The deaths of elephants

African elephants are the world's largest terrestrial mammals: adult males can weigh more than 6,000 kilograms, females average about half as much, and newborns are about 100 kilograms. They are sociable, intelligent, proverbially capacious in memory, and eerily aware of death, as they show in their remarkable behavior when they encounter the bones of their ancestors, lingering at such sites and touching the remnants. Although their bones have remained in Africa, their tusks have often ended up in piano keys or in the ivory bric-a-brac you still sometimes see on mantelpieces.

Ancient Egyptians hunted elephants, and Carthaginians used them in wars with Rome until finally they became extinct in North Africa, remaining abundant only south of the Sahara. The best available estimate of the continent's maximum carrying capacity (including smaller-size forest elephants) was about 27 million animals at the beginning of the 19th century; their actual number might have been closer to 20 million. Today, though, there are well under a million.

Reconstructions of the past ivory trade indicate a fairly steady flow of around 100 tons per year until about 1860, and then a fivefold rise just after 1900. The

Where African elephants still live

trade plunged during the First World War, then rose briefly before another war-induced plunge, after which it resumed its rise, peaking at more than 900 tons a year by the late 1980s. I have integrated these fluctuating harvests and come up with aggregate removals of

55,000 tons of ivory during the 19th century, and at least 40,000 tons during the 20th century.

The latter mass translates into the slaughter of at least 12 million elephants. No good systematic estimates of surviving elephants are available before 1970, and continent-wide estimates indicate steady declines during the closing decades of the 20th century. The Great Elephant Census, a project funded by the late Microsoft cofounder Paul G. Allen, relied on aerial surveys of about 80 percent of the savanna elephant's range. When it was completed in 2016, its final count of 352,271 elephants was 30 percent lower than the best estimate in the mid-1980s.

Other news is deeply discouraging: the number of elephants in Mozambique was halved between 2009 and 2014, to 10,000, and during the same five years more than 85,000 Tanzanian elephants were killed, their total dropping from nearly 110,000 to just 43,000 (the difference is accounted for by an annual 5 percent birth rate). New DNA analyses of large ivory seizures made between 1996 and 2014 have traced some 85 percent of the illegal killing to East Africa, above all in the Selous Game Reserve in southeastern Tanzania, the Niassa Reserve in northern Mozambique, and more recently also in central Tanzania.

Most of the blame has rested with China's continuing demand for ivory, much of which gets turned into elaborately kitschy carvings including statuettes of Mao Zedong, the man responsible for the greatest famine in human history. Recently international pressure finally worked, and China's State Council banned all ivory trade and processing activities as of the end of 2017. This has had some positive effects, but Chinese tourists continue to buy ivory objects when they travel to neighboring countries.

And if the slaughter were to stop, some African regions might face a new problem, evident for years in parts of South Africa: a surfeit of elephants. It is no easy matter to manage expanding numbers of large and potentially destructive animals, especially those living in proximity to farmers and herders.

Why calls for the Anthropocene era may be premature

Many historians and scientists argue that we are living in the Anthropocene, a new epoch characterized by human control of the biosphere. In May 2019, the Anthropocene Working Group formally voted to recognize this new geologic epoch, and its proposal will be considered by the International Commission on Stratigraphy that governs epoch-naming.

My reaction, echoing the Romans: *Festina lente.* Make haste slowly.

To be quite clear, there is no doubt about the pervasiveness of our interference in global biogeochemical cycles and the loss of biodiversity attributable to human actions: the mass dumping of our wastes; the large-scale deforestation and accelerated erosion of soils; the global extent of pollution generated by farming, cities, industries, and transportation. In combination, these man-made impacts are unprecedented, and of a scale that may well imperil the future of our species.

But is our control of the planet's fate really so complete? There is plenty of counterevidence. Fundamental variables that make life on Earth possible—the thermonuclear reactions that power the sun, suffusing the planet with radiation; the planet's shape, rotation, tilt,

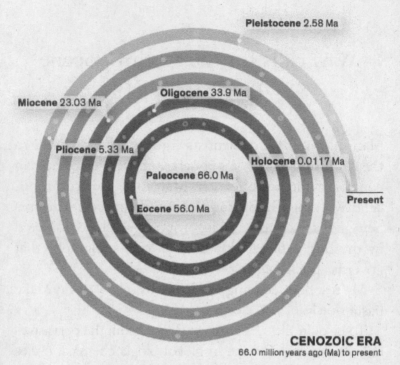

CENOZOIC ERA
66.0 million years ago (Ma) to present

Geologic eras and the Anthropocene

the eccentricity of its orbital path (the "pacemaker" of the ice ages), and the circulation of its atmosphere—are all beyond any human interference. Nor can we ever hope to control the enormous terraforming processes—the Earth's plate tectonics, driven by internal heat and resulting in slow but constant creation of new ocean floor; forming, reshaping, and elevating landmasses whose distributions and altitudes are key determinants of climate variability and habitability.

Similarly, we are mere bystanders watching volcanic eruptions, earthquakes, and tsunamis, the three most violent consequences of plate tectonics. We can live with their frequent, moderate displays, but the very survival of some of the world's largest cities—notably Tokyo, Los Angeles, and Beijing—depends on the absence of mega-earthquakes, and the very existence of modern civilization could be cut short by mega volcanic eruptions. Even when measuring time not in geological but in civilizational terms, we also face far-from-negligible threats from Earth-busting asteroids whose path we might be able to predict but not to alter.

In any given year, these events have very low probabilities, but because of their enormous destructivity their effects are outside the historic human experience. We have no good way to deal with them, but we cannot pretend that, in the long run, they are less relevant than the loss of forest species or the combustion of fossil fuels.

Besides, why rush to elevate ourselves into the creators of a new geological era instead of waiting a bit to see how long the experiment conducted by *Homo sapiens* can last? Each of the six elapsed epochs of the Cenozoic era—from the beginning of the Paleocene 66 million years ago to the beginning of the Holocene 11,700 years ago—lasted at least 2.5 million years, including the previous two (the Pliocene and the Pleistocene), and we are now less than 12,000 years into the Holocene. If there is in fact an Anthropocene, it may date no further back

than 8,000 years (counting since the beginning of settled agriculture) or 150 years (counting from the takeoff
of fossil fuel combustion).

Should we manage to be around for another 10,000
years—a trivial spell for science fiction readers; an eter-
nity for modern, high-energy civilization—we should
congratulate ourselves by naming the era shaped by our
actions. But in the meantime, let us wait before we deter-
mine that our mark on the planet is anything more than
a modest microlayer in the geologic record.

Concrete facts

Ancient Romans invented concrete, a mixture of aggregate (sand, crushed stone), water, and a bonding agent. They called it *opus cementitium*, and this widely used construction material did not contain modern cement (made of lime, clay, and metallic oxides fired in rotating kilns at a high temperature and then ground into fine powder) but rather a mixture of gypsum and quicklime—and its best variety was made with volcanic sand from Puteoli near Mount Vesuvius. Its addition produced an outstanding material fit for massive vaults (Rome's Pantheon, 118–126 CE, remains the world's largest non-reinforced concrete dome) as well as for underwater construction in many ports around the Mediterranean, including ancient Caesarea (located in today's Israel).

Production of modern cement began in 1824, when Joseph Aspdin patented his firing of limestone and clay at high temperatures. Transformation of alumina and silica materials into a non-crystalline amorphous solid (vitrification, the same kind of process that is used to make glass) produces small nodules or lumps of a glassy clinker that is ground to make cement. Cement is then mixed with water (10–15 percent of the final mass) and aggregates (sand and gravel, making up 60–75 percent of

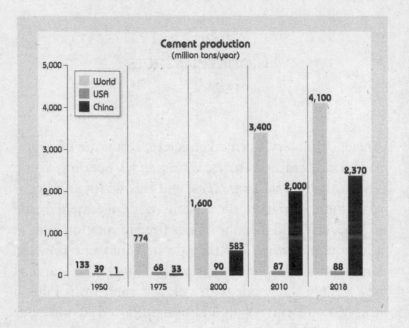

Cement production
(million tons/year)

the total mass) to produce concrete, a formable material
that is strong under compression but weak in tension.

Weakness in tension can be reduced by reinforcement
with steel. The first attempts to do that took place in
France in the early 1860s, but the technique only took
off during the 1880s. The 20th century was the era of
reinforced concrete. In 1903, Cincinnati's Ingalls Building became the world's first reinforced-steel skyscraper;
during the 1930s, structural engineers began to use pre-
stressed concrete (with tensioned steel wires or bars);
and since 1950 the material has been used for buildings
of all heights and functions—the Burj Khalifa tower in
Dubai is the world's tallest, while Jørn Utzon's sail-like

Sydney Opera House is perhaps the material's visually most impressive application. Reinforced concrete made it possible to build massive hydroelectric dams: the world's largest, China's Three Gorges dam, contains three times as much of it as Grand Coulee, America's largest. Concrete bridges are also common: Beipan River Bridge is now the world's longest concrete arch bridge, spanning the 445-meter gorge between two Chinese provinces. But mostly, concrete is deployed in visually unremarkable ways, in the form of billions of railroad sleepers, paved roads, freeways, parking lots, ports, and airport runways and taxiways.

Between 1900 and 1928, the US consumption of cement rose tenfold to 30 million tons, and the postwar economic expansion (including the construction of the Interstate Highway System, which requires about 10,000 tons of concrete per kilometer) raised it to the peak of about 128 million tons by 2005, with the latest rates less than 100 million tons a year.

China became the world's largest producer in 1986, and its output of cement—more than 2.3 billion tons in 2018—now accounts for nearly 60 percent of the global total. The most impressive illustration of China's unprecedented construction effort is that in just the last two years the country emplaced more cement (about 4.7 billion tons) than the US did cumulatively throughout the entire 20th century (about 4.6 billion tons)!

But concrete is not an everlasting material, and the Pantheon's extraordinary longevity is a rare exception.

Concrete deteriorates in all climates, and the process is accelerated by factors ranging from acid deposition to vibration, and from structural overloading to salt-induced corrosion; and in warm and humid environments, algal growth blackens the exposed surfaces. As a result, the post-1950 planet-wide concretization has produced tens of billions of tons of material that will have to be either replaced or destroyed (or simply abandoned) in the coming decades.

The material's environmental impact is another worry. Air pollution (fine dust) from cement production can be captured by fabric filters, but the industry (burning such inferior fuels as low-quality coal and petroleum coke) remains a significant source of carbon dioxide, emitting roughly a ton of the gas per ton of cement. For comparison, producing a ton of steel is associated with emissions of about 1.8 tons of CO_2.

Production of cement now accounts for about 5 percent of global CO_2 emissions from fossil fuels, but its carbon footprint can be lowered by a variety of measures. Old concrete can be recycled, and the crushed material can be reused in construction. Blast furnace slag or fly ash captured in coal-fire power plants can replace some of the cement in mixing concrete. There are also several new low-carbon or no-carbon cement-making processes, but these alternatives will be making only small annual dents in a global output that is now surpassing 4 billion tons.

What's worse for the environment— your car or your phone?

Statistics on the production of energy are fairly reliable; accurate statistics on the consumption of energy by major sectors are harder to come by; and data on the energy consumed in the production of specific goods are even less reliable. Such energy embodied in products is part of the environmental price we pay for everything we own and use.

The estimation of the embodied energy of finished goods relies not only on indisputable facts—so much steel in a car, so many microchips in a computer—but also on the inevitable simplifications and assumptions that must be made to derive overall rates. Which model of car? Which computer or phone? The challenge is to select reasonable, representative rates; the reward is to get a new perspective on the man-made world.

Let's focus on mobile devices and cars. Mobile devices, because they are the primary enablers of instant communication and boundless information; cars, because people still want to move in the real world.

Obviously, a car weighing 1.4 tons (about as much as a Honda Accord LX) embodies more energy than the 140 grams of a smartphone (say, a Samsung Galaxy). But

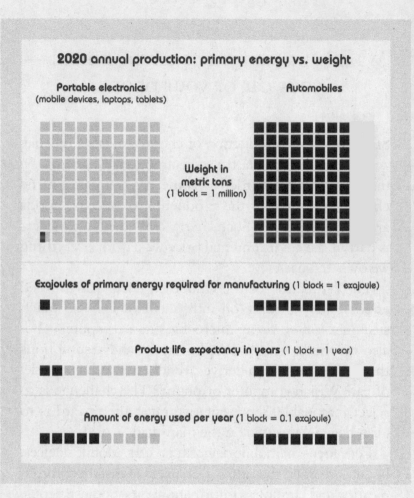

2020 annual production: primary energy vs. weight

Portable electronics
(mobile devices, laptops, tablets)

Automobiles

Weight in
metric tons
(1 block = 1 million)

Exajoules of primary energy required for manufacturing (1 block = 1 exajoule)

Product life expectancy in years (1 block = 1 year)

Amount of energy used per year (1 block = 0.1 exajoule)

the energy gap is nowhere near so great as that 10,000-fold difference in mass.

In 2020, worldwide sales of cellphones should be about 1.75 billion, and those of portable computing devices (laptops, notebooks, tablets) should be on the order of 250 million. The aggregate weight of these devices comes to about 550,000 tons. Assuming, conservatively, an average embodied rate of 0.25 gigajoules per phone, 4.5 gigajoules per laptop, and 1 gigajoule for a tablet, the annual production of these devices requires about 1 exajoule (10^{18} joules) of primary energy—that is, roughly equal to total annual energy use in New Zealand or Hungary. With little less than 100 gigajoules per vehicle, the 75 million vehicles sold in 2020 embody about 7 exajoules of energy (slightly more than Italy's annual energy use) and weigh about 100 million tons. New cars thus weigh more than 180 times as much as all portable electronics, but require only seven times as much energy to make.

And as surprising as that may be, we can make an even more startling comparison. Portable electronics don't last long—on average, just two years—and so the world's annual production of these devices embodies about 0.5 exajoules per year of use. Because passenger cars typically last for at least a decade, the world's annual production embodies about 0.7 exajoules per year of use—which is only 40 percent more than portable electronic devices! I hasten to add that these are, necessarily, only highly approximate calculations—but even if these rough aggregates were to err in opposite directions (that is, if car-making actually embodies more energy than calculated, and electronic production needs less), the global totals would still be surprisingly similar, with the most likely difference no greater than twofold. And looking ahead, the two aggregates might get even closer: annual sales of both cars and mobile devices have been recently slowing, but the future looks less promising for internal combustion engines.

Of course, operating energy costs of the two classes of these highly energy-intensive devices are vastly different. A compact American passenger car consumes about 500 gigajoules of gasoline during a decade of its service, five times its embodied energy cost. A smartphone consumes just 4 kilowatt-hours of electricity annually and less than 30 megajoules during its two years of service—or just 3 percent of its embodied energy cost if the electricity comes from a wind turbine or from a photovoltaic cell. That fraction rises to about 8 percent if the energy comes from burning coal, a less efficient process.

But a smartphone is nothing without a network, and the cost of electrifying the net is high and rising. Forecasts disagree about the coming rate of increase (or about possible stabilization by using innovative designs) but, in any case, those tiny phones leave quite an aggregate footprint in the energy budget—and the environment.

Who has better insulation?

First impressions often lead to wrong conclusions. I well remember receiving a friendly welcome at the residence of a European ambassador in Ottawa, and in the very next sentence being told that this house was perfect to withstand Canadian winters because it was made of real brick and stone—not like those flimsy North American wooden things with hollow walls. My hosts then swiftly moved to other matters, and, in any case, I did not have the heart to belittle the insulating qualities of their handsome home.

The error is easy to understand, but mass and density are better indicators of sturdiness than of insulating capability. A brick wall obviously looks more substantial and protective than a wall framed with narrow wooden studs and covered on the outside with a sheet of thin plywood and aluminum siding and on the inside with vulnerable gypsum drywall. Angry European men do not make holes in brick walls.

Decades ago, when oil sold for $2 a barrel, most pre-1960 North American houses usually had nothing more to keep out the cold than the air space between the plywood and drywall. Sometimes the space was filled with wood shavings or shredded paper. Yet, remarkably, even

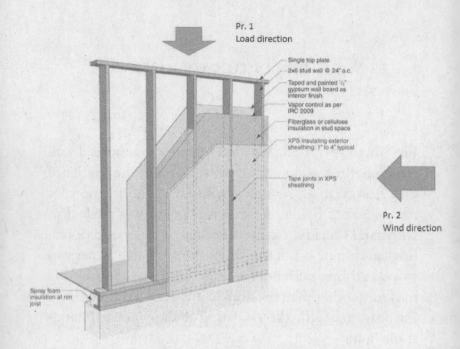

Pr. 1
Load direction

Single top plate
2x6 stud wall @ 24" o.c.
Taped and painted ½" gypsum wall board as interior finish
Vapor control as per IRC 2009
Fiberglass or cellulose insulation in stud space
XPS insulating exterior sheathing: 1" to 4" typical
Tape joints in XPS sheathing

Pr. 2
Wind direction

Spray foam insulation at rim joist

Insulating a wall

that feeble combination provided a bit more insulation than solid brick.

The insulating value, or thermal resistance, is measured in terms of R-value. It depends not only on the composition, thickness, and density of the insulation, but also on the outdoor temperature and moisture. A framed wall from 1960 had roughly the following R-values: aluminum siding (0.6), thin plywood (0.5), air space (0.9), and drywall (0.5). That all adds up to 2.5. But standard brick (0.8) plastered on both sides offered no more than 1.0. Hence even a flimsy, mass-built North American wall insulated at least twice as well as Europe's plastered brick.

Once energy prices began to rise and more rational building codes came into effect in North America, it became compulsory to incorporate plastic barriers and fiberglass batts—pillow-like rolls that can be packed between the wooden frames or studs. Higher overall R-values were easily achieved by using wider studs (two-by-six) or, better yet, by double-studding, which involves building a sandwich from two frames, each one filled with insulation. (In North America, a softwood "two-by-six" is actually 1.5 by 5.5 inches, or 38 by 140 millimeters.) For a well-built North American wall this means adding insulation values of drywall (0.5), polyethylene vapor barrier (0.8), fiberglass batts (20), fiberboard sheathing (1.3), plastic house wrap (Tyvek ThermaWrap at 5), and beveled wood cladding (0.8). Adding the insulating value of interior air film brings the total R-value to about 29.

Brick walls, too, got better. To keep a desired outer look of colored brick, an old wall can be retrofitted from the inside by putting wooden battens (thin strips that hold insulation in place) on the interior plaster and attaching insulation-backed gypsum board integrated with a vapor membrane to keep out moisture. With 2-inch insulated plasterboard, this will triple the previous overall R-value, but even so, the insulated old brick wall will remain an order of magnitude behind the two-by-six framed North American wall. Even people who are generally aware of R-values do not expect to see such a large difference.

However, all this wall insulation can reach its potential only if the windows don't bleed heat (see the next chapter).

Triple-glazed windows: A see-through energy solution

The quest for untested technical fixes is the curse of energy policy. Take your pick: self-driving solar-powered cars, inherently safe nuclear minireactors, or genetically enhanced photosynthesis.

But why not start with what is proven? Why not simply reduce the demand for energy, beginning with residential and commercial buildings?

Both in the United States and in the European Union, buildings account for about 40 percent of total primary energy consumption (transportation comes second, at 28 percent in the US and about 22 percent in the EU). Heating and air conditioning account for half of residential consumption, which is why the single best thing we could do for the energy budget is to keep the heat in (or out) with better insulation.

The most rewarding place to do that is in windows, where the energy loss is the highest. That is to say, it has the highest thermal transmittance, measured in watts passing through a square meter of material, divided by the difference in temperature in kelvins on either side. A single pane has a heat transfer coefficient of 5.7–6 watts per square meter per degree of kelvin; a double pane separated by 6 millimeters (air is a poor conductor of

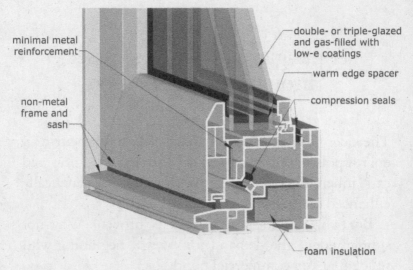

minimal metal
reinforcement

double- or triple-glazed
and gas-filled with
low-e coatings

warm edge spacer

compression seals

non-metal
frame and
sash

foam insulation

Insulating a window

heat) has a coefficient of 3.3. Applying coatings to minimize the passage of ultraviolet and infrared radiation
lowers it to between 1.8 and 2.2, and filling the space
between the panes with argon (in order to slow down
heat transfer) chops it to 1.1. Do that with triple-glazed
windows and you drop to between 0.6 and 0.7. Substitute krypton for argon and you can get it down to 0.5.

That's a loss reduction of up to 90 percent compared
to a single pane. In the world of energy savings, there
are no other opportunities of that magnitude applicable
on a scale of billions of units. Bonus point: it would
actually work.

And there is also a comfort factor. With the outdoor
temperature at −18°C (common January overnight lows

in Edmonton, Alberta, or daily highs in Russia's Novo-sibirsk) and the indoor temperature at 21°C, the internal surface temperature of a single-pane window is around 1°C, an older double-pane window will register 11°C, and the best triple-glazed window 18°C. At that temperature, you can sit right next to it.

And triple-paned windows have the added advantage of reducing condensation on the interior glass by raising its temperature above the dew point. Such windows are already common in Sweden and Norway, but in Canada (with its low-cost natural gas) they may not be mandated before 2030, and as in many other cold-weather jurisdictions, the required standard is still equivalent to just a double-glazed window with one low-emissivity coating.

Cold-weather countries have had a long time to learn about insulation. Not so in the warmer places, which need it now that air conditioning is becoming widespread. Most notably, in rural China and rural India single panes are still the norm. Of course, the temperature differential for hot-weather cooling is not as large as in higher latitudes for heating. For instance, at my home in Manitoba, Canada, the average overnight lows in January are −25°C, enough to make a difference of 40°C even when the thermostat is turned down for the night. On the other hand, air conditioning in many hot and humid regions runs for much longer periods than does heating in Canada or Sweden.

Physics is indisputable, but economics rules. Although triple-glazed windows may cost just 15 percent more than double panes, their payback times are obviously longer, and it is commonly claimed that the step from double to triple design is not justified. That may be so if you ignore improved comfort and reduced window condensation—and, above all, the fact that triple panes will keep reducing energy use for decades to come.

Why, then, do visionaries want to pour money into arcane conversion technologies that may not even work and which, even if they did, would likely have bad side effects on the environment? What's wrong with simple insulation?

Improving the efficiency of household heating

If our climate models are correct, and if we must indeed limit the increase in global warming to 2°C (and preferably to just 1.5°) in order to avoid the serious consequences associated with a planet-wide temperature rise, then we will have to take many unprecedented steps to reduce carbon emissions. Attention commonly focuses on new techniques that result in higher efficiencies—like light-emitting diodes—or that introduce entirely new modes of energy conversion, such as electric cars. Conservation is in principle a more practical solution, but unfortunately (besides, as we have seen, triple-glazed windows) there are few ways to extend it to what has long been the single biggest energy hog in the colder parts of the world: house heating.

About 1.2 billion people need to heat their houses: about 400 million people living in the EU, Ukraine and Russia; another 400 million people in North America living outside the US South and Southwest; and 400 million Chinese in the northeastern, northern, and western regions. And almost everywhere you look, the best available techniques are already as efficient as is practically possible.

It is striking just how rapid the diffusion of efficient

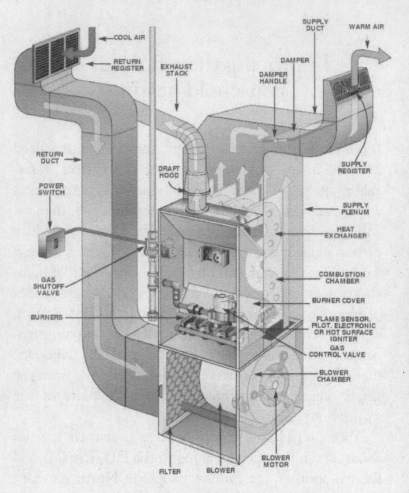

Inside of a household natural gas furnace

systems has been. During the 1950s, my family heated our house near the Czech–German border with wood burned in heavy cast-iron stoves. The efficiency of this process is no higher than 35 percent; the rest of the heat escapes through a chimney. During my studies in Prague in the early 1960s, the city was energized by brown coal—a low-quality lignite—and the stove I stoked had an efficiency of 45 to 50 percent. In the late 1960s, we lived in Pennsylvania on the upper floor of a small suburban house whose old furnace burned oil with about 55 to 60 percent efficiency. In 1973, our first Canadian house had a natural gas furnace rated at 65 percent, and 17 years later, in a newer, superefficient home, I installed a furnace with 94 percent efficiency. I eventually had it replaced with a model rated at 97 percent.

And my progress through a succession of fuels and efficiencies has been replicated by tens of millions of people in the northern hemisphere. Thanks to North America's cheap natural gas and to the combination of (more expensive but readily available) Dutch, North Sea, and Russian gas in Europe, this—the cleanest of all fossil fuels—is what most of the people in northern climates have come to rely on, in place of wood, coal, and fuel oil. In Canada, the production of mid-efficiency (78 to 84 percent) furnaces ended in 2009, and all new houses are now mandated to have high-efficiency (at least 90 percent) furnaces. The same will soon be true elsewhere in the West, while rising gas imports are already causing China to shift its heating from coal to gas.

Future efficiency gains will have to come from somewhere else. Better insulation of the outer-facing part of the house (especially better windows) is the obvious (albeit often expensive) first step. Air-source heat pumps, transferring heat via a heat exchanger, have become popular in many places and are effective as long as temperatures do not dip below freezing; in cold regions they still need winter backup. Solar heating is possible, too, but it doesn't work very well where and when it's needed most—in very cold climates, during prolonged spells of cold but overcast weather, during blizzards, and with the solar modules under heavy snow cover.

Will the long-term need to limit global warming eventually lead to something unthinkable? I am referring to the most economically sensible choice, and the one that would make the greatest, longest-lasting contribution to reducing the carbon burden of heating: limiting the size of houses. We could do away with McMansions—mass-built houses with masses of floor space—in North America. Doing away with similar houses in the tropics would save on the opposite expense—on energy currently wasted on air conditioning. Who's up for that?

Running into carbon

In 1896, Svante Arrhenius, of Sweden, became the first scientist to quantify the effects of man-made carbon dioxide on global temperatures. He calculated that doubling the atmospheric level of the gas from its concentration in his time would raise the average mid-latitude temperature by 5 to 6°C. That's not too far from the latest results, obtained by computer models running more than 200,000 lines of code.

The United Nations held its first Framework Convention on Climate Change in 1992, and this was followed by a series of meetings and climate treaties. But the global emissions of carbon dioxide have been rising steadily just the same.

At the beginning of the 19th century, when the United Kingdom was the only major coal producer, global emissions of carbon from fossil fuel combustion were minuscule, at less than 10 million tons a year (to express them in terms of carbon dioxide, just multiply by 3.66). By century's end, emissions surpassed half a billion tons of carbon. By 1950, they had topped 1.5 billion tons. The postwar economic expansion in Europe, North America, the USSR, and Japan—along with the post-1980 economic rise of China—quadrupled emissions thereafter,

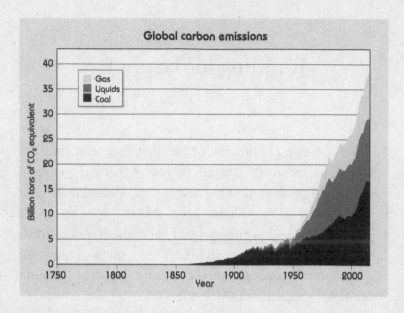

Global carbon emissions

Gas
Liquids
Coal

Billion tons of CO₂ equivalent

Year

to about 7 billion tons of carbon by the year 2000. In the two centuries between 1800 and 2000, the transfer of carbon from fossil fuels to the atmosphere increased 650-fold while the population had increased only sixfold!

The new century has seen a significant divergence. By 2017, emissions had declined by about 15 percent in the European Union, with its slower economic growth and aging population, and also in the United States, thanks largely to the increasing use of natural gas instead of coal. However, all these gains were outbalanced by Chinese carbon emissions, which rose from about 1 billion to about 3 billion tons—enough to increase the worldwide total by nearly 45 percent, to 10.1 billion tons.

By burning huge stocks of carbon that fossilized ages

ago, human beings have pushed carbon dioxide concentrations to levels not seen for about 3 million years. Through drilling deep into glaciers covering Antarctica and Greenland, we can recover slender tubes of ice that contain tiny bubbles, and as we drill deeper, increasingly older ice. By sampling the air locked in those tiny bubbles, we have been able to reconstruct the history of carbon dioxide concentrations going back some 800,000 years. Back then, the atmospheric levels of the gas fluctuated between 180 and 280 parts per million (that is, from 0.018 to 0.028 percent). During the past millennium, the concentrations remained fairly stable, ranging from 275 ppm in the early 1600s to about 285 ppm before the end of the 19th century. Continuous measurements of the gas began near the top of Mauna Loa, in Hawaii, in 1958: the 1959 mean was 316 ppm, the 2015 average reached 400 ppm, and 415 ppm was first recorded in May 2019.

Emissions will continue to decline in affluent countries, and the rate at which they grow in China has begun to slow down. However, it is speeding up in India and Africa, and hence it is unlikely that we will see any substantial global declines anytime soon.

The Paris Agreement of 2015 was lauded as the first accord containing specific national commitments to reduce future emissions. But actually, only a small number of countries made specific promises, there is no binding enforcement mechanism, and even if all those targets were met by 2030, carbon emissions would still rise to nearly 50 percent above the 2017 level. According to the 2018 study by the Intergovernmental Panel for Climate Change, the only way to keep the average world temperature rise to no more than 1.5°C would be to put emissions almost immediately into a decline steep enough to bring them to zero by 2050.

That is not impossible—but it is very unlikely. Reaching that goal would require nothing short of a fundamental transformation of the global economy on scales and at a speed unprecedented in human history, a task that would be impossible to do without major economic and social dislocations. The greatest challenge would be how to lift billions from poverty without relying on fossil carbon. The affluent world has used hundreds of billions of tons of it to create its high quality of life, but right now we do not have any affordable non-carbon alternatives that could be rapidly deployed on mass scales in order to energize the production of enormous quantities of what I have called the four pillars of modern civilization—ammonia, steel, cement, and plastics—which will be needed in Africa and Asia in the decades to come. The contrasts between the expressed concerns about global warming, the continued release of record volumes of carbon, and our capabilities to change that in the near term could not be starker.

Epilogue

Numbers may not lie, but which truth do they convey? In this book I have tried to show that we often have to look both deeper and wider. Even fairly reliable— indeed, even *impeccably accurate*—numbers need to be seen in wider contexts. An informed judging of absolute values requires some relative, comparative perspectives.

Rigid ranking based on minuscule differences misleads rather than informs. Rounding and approximation is superior to unwarranted and unnecessary precision. Doubt, caution, and incessant questioning are in order— but so is the insistence on quantifying the complex realities of the modern world. If we are to understand many unruly realities, if we are to base our decisions on the best available information, then there is no substitute for this pursuit.

Further Reading

People—The Inhabitants of Our World

What happens when we have fewer children?

Bulatao, R.A. and J.B. Casterline, eds. *Global Fertility Transition*. New York: Population Council, 2001.

United Nations. *World Population Prospects*. New York: United Nations, 2019. https://population.un.org/wpp/.

The best indicator of quality of life? Try infant mortality

Bideau, A., B. Desjardins, and H.P. Brignoli, eds. *Infant and Child Mortality in the Past*. Oxford: Clarendon Press, 1992.

Galley, C., et al., eds. *Infant Mortality: A Continuing Social Problem*. London: Routledge, 2017.

The best return on investment: Vaccination

Gates, Bill and Melinda. "Warren Buffett's Best Investment." *Gates Notes* (blog), February 14, 2017. https://www. gatesnotes.com/2017-Annual-Letter?WT.mc_id=02_14_ 2017_02_AL2017GFO_GF-GFO_&WT.tsrc=GFGFO.

Ozawa, S., et al. "Modeling the economic burden of adult vaccine-preventable diseases in the United States." *Health Affairs* 35, no. 11 (2016): 2124–32.

Why it's difficult to predict how bad a pandemic
will be while it is happening

NHCPRC (National Health Commission of the People's Republic of China). "March 29: Daily briefing on novel coronavirus cases in China." March 29, 2020. http://en.nhc.gov.cn/2020-03/29/c_78447.htm.

Wong, J.Y., et al. "Case fatality risk of influenza A (H1N1pdm09): A systematic review." *Epidemiology* 24, no. 6 (2013). https://doi.org/10.1097/EDE.0b013e3182a67448.

Growing taller

Floud, R. et al. *The Changing Body*. Cambridge: Cambridge University Press, 2011.

Koletzko, B., et al., eds. *Nutrition and Growth: Yearbook 2018*. Basel: Karger, 2018.

Is life expectancy finally topping out?

Riley, J.C. *Rising Life Expectancy: A Global History*. Cambridge: Cambridge University Press, 2001.

Robert, L., et al. "Rapid increase in human life expectancy: Will it soon be limited by the aging of elastin?" *Biogerontology* 9, no. 2 (April 2008): 119–33.

How sweating improved hunting

Jablonski, N.G. "The naked truth." *Scientific American Special Editions* 22, 1s (December 2012). https://doi.org/10.1038/scientificamericanhuman1112-22.

Taylor, N.A.S., and C.A. Machado-Moreira. "Regional variations in transepidermal water loss, eccrine sweat gland density, sweat secretion rates and electrolyte composition in resting and exercising humans." *Extreme Physiology and Medicine* 2, no. 4 (2013). https://doi.org/10.1186/2046-7648-2-4.

How many people did it take to build the Great Pyramid?

Lehner, M. *The Complete Pyramids: Solving the Ancient Mysteries.* London: Thames and Hudson, 1997.
Mendelssohn, K. *The Riddle of the Pyramids.* London: Thames and Hudson, 1974.

Why unemployment figures do not tell the whole story

Knight, K.G. *Unemployment: An Economic Analysis.* London: Routledge, 2018.
Summers, L.H., ed. *Understanding Unemployment.* Cambridge, MA: MIT Press, 1990.

What makes people happy?

Heliwell, J.F., R. Layard, and J.D. Sachs, eds. *World Happiness Report 2019.* New York: Sustainable Development Solutions Network, 2019. https://s3.amazonaws.com/happiness-report/2019/WHR19.pdf
Layard, R. *Happiness: Lessons from a New Science.* London: Penguin Books, 2005.

The rise of megacities

Canton, J. 2011. "The extreme future of megacities." *Signifi-cance* 8, no. 2 (June 2011): 53–6. https://doi.org/10.1111/j.1740-9713.2011.00485.x.

Munich Re. *Megacities—Megarisks: Trends and challenges for insurance and risk management.* Munich: MünchenerRück versicherungs-Gesellschaft, 2004. http://www.prevention web.net/files/646_10363.pdf.

Countries—Nations in the Age of Globalization

The First World War's extended tragedies

Bishop, C., ed. *The Illustrated Encyclopedia of Weapons of World War I.* New York: Sterling Publishing, 2014.

Stoltzenberg, D. *Fritz Haber: Chemist, Nobel Laureate, German, Jew.* Philadelphia, PA: Chemical Heritage Foundation, 2004.

Is the US really exceptional?

Gilligan, T.W., ed. *American Exceptionalism in a New Era: Rebuilding the Foundation of Freedom and Prosperity.* Stanford, CA: Hoover Institution Press, 2018.

Hodgson, G. *The Myth of American Exceptionalism.* New Haven, CT: Yale University Press, 2009.

Why Europe should be more pleased with itself

Bootle, R. *The Trouble with Europe: Why the EU Isn't Working, How It Can Be Reformed, What Could Take Its Place.* Boston, MA: Nicholas Brealey, 2016.

Leonard, D., and M. Leonard, eds. *The Pro-European Reader.* London: Palgrave/Foreign Policy Centre, 2002.

Brexit: Realities that matter most will not change

Clarke, H.D., M. Goodwin, and P. Whiteley. *Brexit: Why Britain Voted to Leave the European Union.* Cambridge: Cambridge University Press, 2017.

Merritt, G. *Slippery Slope: Brexit and Europe's Troubled Future.* Oxford: Oxford University Press, 2017.

Concerns about Japan's future

Cannon, M.E., M. Kudlyak, and M. Reed. "Aging and the economy: The Japanese experience." *Regional Economist* (October 2015). https://www.stlouisfed.org/publications/regional-economist/october-2015/aging-and-the-economy-the-japanese-experience.

Glosserman, B. *Peak Japan: The End of Great Ambitions.* Washington, DC: Georgetown University Press, 2019.

How far can China go?

Dotsey, M., W. Li, and F. Yang. "Demographic aging, industrial policy, and Chinese economic growth." Federal

Reserve Bank of Philadelphia. *Working Papers* (2019): 19–21. https://doi.org/10.21799/frbp.wp.2019.21.

Paulson Jr., H.M. *Dealing with China: An Insider Unmasks the New Economic Superpower.* New York: Twelve, 2016.

India vs. China

Drèze, J., and A. Sen. *An Uncertain Glory: India and Its Contradictions.* Princeton, NJ: Princeton University Press, 2015.

NITI Aayog. *Strategy for New India @ 75.* November 2018. https://niti.gov.in/writereaddata/files/Strategy_for_New_India.pdf.

Why manufacturing remains important

Haraguchi, N., C.F.C. Cheng, and E. Smeets. "The importance of manufacturing in economic development: Has this changed?" Inclusive and Sustainable Development Working Paper Series WP1, 2016. https://www.unido.org/sites/default/files/2017-02/the_importance_of_manufacturing_in_economic_development_0.pdf.

Smil, V. *Made in the USA: The Rise and Retreat of American Manufacturing.* Cambridge, MA: MIT Press, 2013.

Russia and the USA: How things never change

Divine, R.A. *The Sputnik Challenge: Eisenhower's Response to the Soviet Satellite.* Oxford: Oxford University Press, 2003.

Zarya. "Sputniks into Orbit." http://www.zarya.info/Diaries/Sputnik/Sputnik1.php.

Receding empires: Nothing new under the sun

Arbesman, S. "The life-spans of empires." *Historical Methods* 44, no. 3 (2011): 127–9. https://doi.org/10.1080/01615440. 2011.577733.

Smil, V. *Growth: From Microorganisms to Megacities.* Cambridge, MA: MIT Press, 2019.

Machines, Designs, Devices—Inventions That Made Our Modern World

How the 1880s created our modern world

Smil, V. *Creating the Twentieth Century: Technical Innovations of 1867–1914 and Their Lasting Impact.* Oxford: Oxford University Press, 2005.

Timmons, T. *Science and Technology in Nineteenth-Century America.* Westport, CT: Greenwood Press, 2005.

How electric motors power modern civilization

Cheney, M. *Tesla: Man Out of Time.* New York: Dorset Press, 1981.

Hughes, A. *Electric Motors and Drives: Fundamentals, Types and Applications.* Oxford: Elsevier, 2005

Transformers—the unsung silent, passive devices

Coltman, J.W. "The transformer." *Scientific American* 258, no. 1 (January 1988): 86–95.

Harlow, J.H., ed. *Electric Power Transformer Engineering.* Boca Raton, FL: CRC Press, 2012.

Why you shouldn't write diesel off just yet

Mollenhauer, K., and H. Tschöke, eds. *Handbook of Diesel Engines.* Berlin: Springer, 2010.

Smil, V. *Prime Movers of Globalization: The History and Impact of Diesel Engines and Gas Turbines.* Cambridge, MA: MIT Press, 2010.

Capturing motion—from horses to electrons

Eadweard Muybridge Online Archive. "Galleries." http://www.muybridge.org/.

Muybridge, E. *Descriptive Zoopraxography, or the Science of Animal Locomotion Made Popular.* Philadelphia, PA: University of Pennsylvania, 1893. https://archives.upenn.edu/digitized-resources/docs-pubs/muybridge/descriptive-zoopraxography.

From the phonograph to streaming

Marco, G. A., ed. *Encyclopedia of Recorded Sound in the United States.* New York: Garland Publishing, 1993.

Morris, E. *Edison.* New York: Random House, 2019.

Inventing integrated circuits

Berlin, L. *The Man Behind the Microchip: Robert Noyce and the Invention of Silicon Valley.* Oxford: Oxford University Press, 2006.

Lécuyer, C., and D.C. Brook. *Makers of the Microchip: A Documentary History of Fairchild Semiconductor.* Cambridge, MA: MIT Press, 2010.

Moore's Curse: Why technical progress takes longer than you think

Mody, C.C.M. *The Long Arm of Moore's Law: Microelectronics and American Science.* Cambridge, MA: MIT Press, 2016.
Smil, V. *Growth: From Microorganisms to Megacities.* Cambridge, MA: MIT Press, 2019.

The rise of data: Too much too fast

Hilbert, M., and P. López. "The world's technological capacity to store, communicate, and compute information." *Science* 332, no. 6025 (April 2011): 60–65. https://doi.org/0.116/science.1200976.
Reinsel, D., J. Gantz, and J. Rydning. *Data Age 2025: The Digitization of the World: From Edge to Core.* Seagate, 2017. https://www.seagate.com/files/www-content/our-story/trends/files/Seagate-WP-DataAge2025-March-2017.pdf.

Being realistic about innovation

Schiffer, M.B. *Spectacular Failures: Game-Changing Technologies that Failed.* Clinton Corners, NY: Eliot Werner Publications, 2019.
Smil, V. *Transforming the Twentieth Century.* Oxford: Oxford University Press, 2006.

Fuels and Electricity—Energizing Our Societies

Why gas turbines are the best choice

American Society of Mechanical Engineers. *The World's First Industrial Gas Turbine Set—GT Neuchâtel: A Historical Mechanical Engineering Landmark.* Alstom, 1988. https://www.asme.org/wwwasmeorg/media/resourcefiles/aboutasme/who%20we%20are/engineering%20history/landmarks/135-neuchatel-gas-turbine.pdf.

Smil, V. *Natural Gas: Fuel for the Twenty-First Century.* Chichester, West Sussex: John Wiley, 2015.

Nuclear electricity—an unfulfilled promise

International Atomic Energy Agency. *Nuclear Power Reactors in the World.* Reference Data Series No. 2. Vienna: IAEA, 2019. https://www-pub.iaea.org/MTCD/Publications/PDF/RDS-2-39_web.pdf.

Smil, V. *Energy and Civilization: A History.* Cambridge, MA: MIT Press, 2017.

Why you need fossil fuels to get electricity from wind

Ginley, D.S., and D. Cahen, eds. *Fundamentals of Materials for Energy and Environmental Sustainability.* Cambridge: Cambridge University Press, 2012.

Mishnaevsky Jr., L., et al. "Materials for wind turbine blades: An overview." *Materials* 10 (2017). https://www.ncbi.nlm. nih.gov/pmc/articles/PMC5706232/pdf/materials-10-01285. pdf.

How big can a wind turbine be?

Beurskens, J. "Achieving the 20 MW Wind Turbine." *Renewable Energy World* 1, no. 3 (2019). https://www.renewable energyworld.com/articles/print/special-supplement-wind-technology/volume-1/issue-3/wind-power/achieving-the-20-mw-wind-turbine.html.

General Electric. "Haliade-X 12 MW offshore wind turbine platform." Accessed December 2019. https://www.ge.com/renewableenergy/wind-energy/offshore-wind/haliade-x-offshore-turbine.

The slow rise of photovoltaics

NASA. "Vanguard 1." Accessed December 2019. https://nssdc. gsfc.nasa.gov/nmc/spacecraft/display.action?id=1958-002B.

US Department of Energy. "The History of Solar." Accessed December 2019. https://www1.eere.energy.gov/solar/pdfs/solar_timeline.pdf.

Why sunlight is still best

Arecchi, A.V., T. Messadi, and R.J. Koshel. *Field Guide to Illumination.* SPIE, 2007. https://doi.org/10.1117/3.764682.

Pattison, P.M., M. Hansen, and J.Y. Tsao. "LED lighting efficacy: Status and directions." *Comptes Rendus* 19, no. 3 (2017). https://www.osti.gov/pages/servlets/purl/1421610.

Why we need bigger batteries

Korthauer, R., ed. *Lithium-Ion Batteries: Basics and Applications.* Berlin: Springer, 2018.

Wu, F., B. Yang, and J. Ye, eds. *Grid-Scale Energy Storage Systems and Applications.* London: Academic Press, 2019.

Why electric container ships are a hard sail

Kongsberg Maritime. "Autonomous Ship Project, Key Facts about *Yara Birkeland*." Accessed December 2019. https://www.kongsberg.com/maritime/support/themes/autonomous-ship-project-key-facts-about-yara-birkeland/.

Smil, V. *Prime Movers of Globalization: The History and Impact of Diesel Engines and Gas Turbines.* Cambridge, MA: MIT Press, 2010.

The real cost of electricity

Eurostat. "Electricity price statistics." Data extracted November 2019. https://ec.europa.eu/eurostat/statistics-explained/pdfscache/45239.pdf.

Vogt, L.J. *Electricity Pricing: Engineering Principles and Methodologies.* Boca Raton, FL: CRC Press, 2009.

The inevitably slow pace of energy transitions

International Energy Agency. *World Energy Outlook 2019*. Paris: IEA, 2019. https://www.iea.org/reports/world-energy-outlook-2019.

Smil, V. *Energy Transitions: Global and National Perspectives*. Santa Barbara, CA: Praeger, 2017.

Transport—How We Get Around

Shrinking the journey across the Atlantic

Griffiths, D. *Brunel's Great Western*. New York: Harper-Collins, 1996.

Newall, P. *Ocean Liners: An Illustrated History*. Barnsley, South Yorkshire: Seaforth Publishing, 2018.

Engines are older than bicycles!

Bijker, W.E. *Of Bicycles, Bakelites and Bulbs: Toward a Theory of Sociotechnical Change*. Cambridge, MA: MIT Press, 1995.

Wilson, D.G. *Bicycling Science*. Cambridge, MA: MIT Press, 2004.

The surprising story of inflatable tires

Automotive Hall of Fame. "John Dunlop." Accessed December 2019. https://www.automotivehalloffame.org/honoree/john-dunlop/.

Dunlop, J.B. *The History of the Pneumatic Tyre*. Dublin: A. Thom & Co., 1925.

When did the age of the car begin?

Casey, R.H. *The Model T: A Centennial History.* Baltimore, MD: Johns Hopkins University Press, 2008.

Ford Motor Company. "Our History—Company Timeline." Accessed December 2019. https://corporate.ford.com/history.html.

Modern cars have a terrible weight-to-payload ratio

Lotus Engineering. *Vehicle Mass Reduction Opportunities.* October 2010. https://www.epa.gov/sites/production/files/2015-01/documents/10052010mstrs_peterson.pdf.

US Environmental Protection Agency. *The 2018 EPA Automotive Trends Report: Greenhouse Gas Emissions, Fuel Economy, and Technology since 1975.* Executive summary, 2019. https://nepis.epa.gov/Exe/ZyPDF.cgi?Dockey=P100W3WO.pdf.

Why electric cars aren't as great as we think (yet)

Deloitte. *New Market. New Entrants. New Challenges: Battery Electric Vehicles.* 2019. https://www2.deloitte.com/content/dam/Deloitte/uk/Documents/manufacturing/deloitte-uk-battery-electric-vehicles.pdf.

Qiao, Q., et al. "Comparative study on life cycle CO_2 emissions from the production of electric and conventional cars in China." *Energy Procedia* 105 (2017): 3584–95.

When did the jet age begin?

Smil, V. *Prime Movers of Globalization: The History and Impact of Diesel Engines and Gas Turbines.* Cambridge, MA: MIT Press, 2009.

Yenne, B. *The Story of the Boeing Company.* London: Zenith Press, 2010.

Why kerosene is king

CSA B836. *Storage, Handling, and Dispensing of Aviation Fuels at Aerodromes.* Toronto: CSA Group, 2014.

Vertz, L., and S. Sayal. "Jet fuel demand flies high, but some clouds on the horizon." *Insight* 57 (January 2018). https://cdn.ihs.com/www/pdf/Long-Term-Jet-Fuel-Outlook-2018.pdf.

How safe is flying?

Boeing. *Statistical Summary of Commercial Jet Airplane Accidents: Worldwide Operations 1959–2017.* Seattle, WA: Boeing Commercial Airplanes, 2017. https://www.boeing.com/resources/boeingdotcom/company/about_bca/pdf/statsum.pdf.

International Civil Aviation Organization. *State of Global Aviation Safety.* Montreal: ICAO, 2019. https://www.icao.int/safety/Documents/ICAO_SR_2019_29082019.pdf.

Which is more energy efficient—planes, trains, or automobiles?

Davis, S.C., S.W. Diegel, and R.G. Boundy. *Transportation Energy Data Book.* Oak Ridge, TN: Oak Ridge National

Laboratory, 2019. https://info.ornl.gov/sites/publications/files/Pub31202.pdf.

Sperling, D., and N. Lutsey. "Energy efficiency in passenger transportation." *Bridge* 39, no. 2 (Summer 2009): 22–30.

Food—Energizing Ourselves

The world without synthetic ammonia

Smil, V. *Enriching the Earth: Fritz Haber, Carl Bosch, and the Transformation of World Food Production.* Cambridge, MA: MIT Press, 2000.

Stoltzenberg, D. *Fritz Haber: Chemist, Nobel Laureate, German, Jew.* Philadelphia, PA: Chemical Heritage Foundation, 2004.

Multiplying wheat yields

Calderini, D.F., and G.A. Slafer. "Changes in yield and yield stability in wheat during the 20th century." *Field Crops Research* 57, no. 3 (1998): 335–47.

Smil, V. *Growth: From Microorganisms to Megacities.* Cambridge, MA: MIT Press, 2019.

The inexcusable magnitude of global food waste

Gustavsson, J., et al. *Global Food Losses and Food Waste.* Rome: Food and Agriculture Organization of the United Nations, 2011.

WRAP. *The Food Waste Reduction Roadmap—Progress Report 2019*. September 2019. http://wrap.org.uk/sites/files/wrap/Food-Waste-Reduction_Roadmap_Progress-Report-2019.pdf.

The slow addio *to the Mediterranean diet*

Tanaka, T., et al. "Adherence to a Mediterranean diet protects from cognitive decline in the invecchiare in Chianti study of aging. *Nutrients* 10, no. 12 (2007). https://doi.org/10.3390/nu10122007.

Wright, C.A. *A Mediterranean Feast: The Story of the Birth of the Celebrated Cuisines of the Mediterranean, from the Merchants of Venice to the Barbary Corsairs.* New York: William Morrow, 1999.

Bluefin tuna: On the way to extinction

MacKenzie, B.R., H. Mosegaard, and A.A. Rosenberg. "Impending collapse of bluefin tuna in the northeast Atlantic and Mediterranean." *Conservation Letters* 2 (2009): 25–34.

Polacheck, T., and C. Davies. *Considerations of Implications of Large Unreported Catches of Southern Bluefin Tuna for Assessments of Tropical Tunas, and the Need for Independent Verification of Catch and Effort Statistics.* CSIRO Marine and Atmospheric Research Paper No. 23, March 2008. http://www.iotc.org/files/proceedings/2008/wptt/IOTC-2008-WPTT-INF01.pdf.

Why chicken rules

National Chicken Council. "U.S. Broiler Performance." Updated March 2019. https://www.nationalchickencouncil. org/about-the-industry/statistics/u-s-broiler-performance/.

Smil, V. *Should We Eat Meat?: Evolution and Consequences of Modern Carnivory.* Chichester, West Sussex: Wiley-Blackwell, 2013.

(Not) drinking wine

Aurand, J.-M. *State of the Vitiviniculture World Market.* International Organization of Vine and Wine, 2018. http://www.oiv.int/public/medias/6370/state-of-the-world-vitiviniculture-oiv-2018-ppt.pdf.

Lejeune, D. *Boire et Manger en France, de 1870 au Début des Années 1990.* Paris: Lycée Louis le Grand, 2013.

Rational meat-eating

Pereira, P., et al. "Meat nutritional composition and nutritive role in the human diet." *Meat Science* 93, no. 3 (March 2013): 589–92. https://doi.org/10.1016/j.meatsci.2012.09.018.

Smil, V. *Should We Eat Meat?: Evolution and Consequences of Modern Carnivory.* Chichester, West Sussex: Wiley-Blackwell, 2013.

The Japanese diet

Cwiertka, K.J. *Modern Japanese Cuisine: Food, Power and National Identity.* London: Reaktion Books, 2006.

Smil, V., and K. Kobayshi. *Japan's Dietary Transition and Its Impacts*. Cambridge, MA: MIT Press, 2012.

Dairy products—the counter-trends

American Farm Bureau Federation. "Trends in beverage milk consumption." Market Intel, December 19, 2017. https://www.fb.org/market-intel/trends-in-beverage-milk-consumption.

Watson, R.R., R.J. Collier, and V.R. Preedy, eds. *Nutrients in Dairy and Their Implications for Health and Disease*. London: Academic Press, 2017.

Environment—Damaging and Protecting Our World

Animals vs. artifacts—which are more diverse?

GSMArena. "All mobile phone brands." Accessed December 2019. https://www.gsmarena.com/makers.php3.

Mora, C., et al. "How many species are there on Earth and in the ocean?" *PLoS Biology* 9, no. 8 (2011): e1001127. https://doi.org/10.1371/journal.pbio.1001127.

Planet of the cows

Beef Cattle Research Council. "Environmental Footprint of Beef Production." Updated October 23, 2019. https://www.beefresearch.ca/research-topic.cfm/environmental-6.

Smil, V. *Harvesting the Biosphere: What We Have Taken from Nature*. Cambridge, MA: MIT Press, 2013.

The deaths of elephants

Paul G. Allen Project. *The Great Elephant Census Report 2016.* Vulcan Inc., 2016. http://www.greatelephantcensus.com/final-report.

Pinnock, D., and C. Bell. *The Last Elephants.* London: Penguin Random House, 2019.

Why calls for the Anthropocene era may be premature

Davies, J. *The Birth of the Anthropocene.* Berkeley, CA: University of California Press, 2016.

Subcommission on Quaternary Stratigraphy, "Working Group on the 'Anthropocene.'" May 21, 2019. http://quaternary.stratigraphy.org/working-groups/anthropocene/.

Concrete facts

Courland, R. *Concrete Planet: The Strange and Fascinating Story of the World's Most Common Man-Made Material.* Amherst, NY: Prometheus Books, 2011.

Smil, V. *Making the Modern World: Materials and Dematerialization.* Chichester, West Sussex: John Wiley and Sons, 2014.

What's worse for the environment—your car or your phone?

Anders, S.G., and O. Andersen. "Life cycle assessments of consumer electronics—are they consistent?" *International Journal of Life Cycle Assessment* 15 (July 2010): 827–36.

Qiao, Q., et al. "Comparative study on life cycle CO_2 emissions from the production of electric and conventional cars in China." *Energy Procedia* 105 (2017): 3584–95.

Who has better insulation?

Natural Resources Canada. *Keeping the Heat In.* Ottawa: Energy Publications, 2012. https://www.nrcan.gc.ca/energy-efficiency/energy-efficiency-homes/how-can-i-make-my-home-more-ener/keeping-heat/15768.

US Department of Energy. "Insulation materials." Accessed December 2019. https://www.energy.gov/energysaver/weatherize/insulation/insulation-materials.

Triple-glazed windows: A see-through energy solution

Carmody, J., et al. *Residential Windows: A Guide to New Technology and Energy Performance.* New York: W.W. Norton and Co., 2007.

US Department of Energy. *Selecting Windows for Energy Efficiency.* Merrifield, VA: Office of Energy Efficiency, 2018. https://nascsp.org/wp-content/uploads/2018/02/us-doe_selecting-windows-for-energy-efficiency.pdf.

Improving the efficiency of household heating

Energy Solutions Center. "Natural gas furnaces." December 2008. https://naturalgasefficiency.org/for-residential-customers/heat-gas_furnace/.

Lechner, N. *Heating, Cooling, Lighting.* Hoboken, NJ: John Wiley and Sons, 2014.

Running into carbon

Jackson, R.B., et al. *Global Energy Growth Is Outpacing Decarbonization.* A special report for the United Nations Climate Action Summit, September 2019. Canberra: Global Carbon Project, 2019. https://www.globalcarbonproject.org/global/pdf/GCP_2019_Global%20energy%20growth%20outpace%20decarbonization_UN%20Climate%20Summit_HR.pdf.

Smil, V. *Energy Transitions: Global and National Perspectives.* Santa Barbara, CA: Praeger, 2017.

Acknowledgments

For many years, as I kept publishing interdisciplinary books, I thought that it might be an interesting challenge to have a regular opportunity to comment on some newsworthy topics, to straighten some common misconceptions, and to explain some fascinating realities of the modern world. I also thought that the likelihood of ever doing that was fairly low, because in order to be worth doing, an offer from a publisher would have to meet several Goldilocks criteria.

The interval between the contributions should be neither too short (weekly would be a chore) nor too sporadic. The word quota not too long, but long enough to allow for more than a few shallow paragraphs. The pitch neither too specialized nor too superficial, in order to permit an informed examination. Choice of topics not unlimited (I had no intention of writing about obscure matters or exceedingly specialized themes) but definitely wide-ranging. And a tolerance of numbers: not too many, but enough to make a convincing case. The last point was particularly important to me, because over the decades I had noticed how the discussion of important matters that demand some clear quantitative understanding was becoming increasingly qualitative and hence progressively unmoored from complex realities.

Unlikely things happen—and in 2014 I was asked to write a monthly essay for *IEEE Spectrum*, the magazine of the

Institute of Electrical and Electronics Engineers, which is headquartered in New York. Philip Ross, a senior editor at *Spectrum*, proposed my name, and Susan Hassler, the editor-in-chief, readily concurred. *Spectrum* is the flagship magazine (and website) of the world's largest professional organization devoted to engineering and the applied sciences, and its members have been at the forefront of transforming a modern world dependent on the incessant, affordable, and reliable supply of electricity and on the adoption of a growing array of new electronic devices and computerized solutions.

In an email to Phil in October 2014, I outlined the topics intended for the first year. They ranged from cars that weigh too much to triple windows and from Moore's Curse to the Anthropocene. Nearly the entire original line-up was eventually written and printed, starting in January 2015, with the first monthly column on ever-heavier cars. *Spectrum* has been the perfect home for my essays. With its more than 400,000 members IEEE provides a large, highly educated and critical readership, I have been given complete freedom to select the topics, and Phil has been an exemplary editor, particularly relentless in his fact-checking pursuits.

As the essays accumulated, I thought they might make an interesting collection but, again, I did not see much chance of seeing them in a book form. And then in late October 2019, almost exactly five years after I outlined the first year's line-up of essays to Phil, came another unexpected email from Daniel Crewe, Publisher at Viking (Penguin Random House) in London, who was wondering if I had considered turning my columns into a book. Everything then moved very quickly. Daniel secured permissions from Susan, we chose three scores of published essays for the collection (leaving out only a few highly technical columns), and I wrote a new dozen in order to round out the seven topical chapters (especially on food and people). Connor Brown did the first major edit, and we selected appropriate graphics and photographs.

Thanks to Phil and Susan and to the readers of *Spectrum* for the support and for the opportunity to write about anything that captures my curiosity, and to Daniel and Connor for giving these quantitative musings the second life.

The majority of the illustrations come from private collections. Others are from:

p. 98, The miraculous 1880s © Erik Vrielink; p. 105, The world's largest transformer: Siemens for China © Siemens; p. 151, Comparisons of wind turbine heights and blade diameters © Chao (Chris) Qin; p. 155, An aerial view of the Ouarzazate Noor Power Station in Morocco. At 510 MW, it is the world's largest central solar power and photovoltaic installation © Fadel Senna via Getty; p. 167, Model of *Yara Birkeland* © Kongsberg; p. 240, Another record price for a bluefin tuna © Reuters, Kim Kyung-Hoon; p. 276, Where African elephants still live © Vulcan Inc.; p. 280, Geologic eras and the Anthropocene © Erik Vrielink.

ACKNOWLEDGMENTS

First published as . . .

The best return on investment: Vaccination 12
Vaccination: The Best Return on Investment (2017)

Is life expectancy finally topping out? 24
Is Life Expectancy Finally Topping Out? (2019)

How sweating improved hunting 28
The Energy Balance of Running (2016)

How many people did it take to build the Great Pyramid? 31
Building the Great Pyramid (2020)

Why unemployment figures do not tell the whole story 35
Unemployment: Pick a Number (2017)

The First World War's extended tragedies 53
November 1918: The First World War Ends (2018)

Is the US really exceptional? 57
American Exceptionalism (2015)

Why Europe should be more pleased with itself 61
January 1958: European Economic Community (2018)

Concerns about Japan's future 69
'New Japan' at 70 (2015)

When did the jet age begin? 204
October 1958: First Boeing 707 to Paris (2018)

Why kerosene is king 208
Flying Without Kerosene (2016)

Which is more energy efficient—planes, trains, or
   automobiles? 216
Energy Intensity of Passenger Travel (2019)

The inexcusable magnitude of global food waste 230
Food Waste (2016)

The slow *addio* to the Mediterranean diet 235
Addio to the Mediterranean Diet (2016)

Bluefin tuna: On the way to extinction 239
Bluefin Tuna: Fast, but Maybe Not Fast Enough (2017)

Why chicken rules 243
Why Chicken Rules (2020)

(Not) drinking wine 247
(Not) Drinking Wine (2020)

Animals vs. artifacts—which are more diverse? 267
Animals vs. Artifacts: Which are more diverse? (2019)

Planet of the cows 271
Planet of the Cows (2017)

Index

Numbers in *italics* refer to maps, tables, graphs and illustrations.

VACLAV SMIL

HOW THE WORLD REALLY WORKS

We have never had so much information at our fingertips and yet most of us simply don't understand how our world really works. Professor Vaclav Smil is not a pessimist or an optimist, he is a scientist, and this book is a much-needed reality check on topics ranging from food production and nutrition, through energy and the environment, to globalization and the future. For example, the carbon footprint of meat is well known, but did you know that the equivalent of five tablespoons of diesel fuel goes into the production of *each* greenhouse-grown, medium-size, supermarket-bought tomato? The gap between belief and reality is vast.

Drawing on the latest science, tackling sources of misinformation head on and championing a rational, fact-based approach, in *How the World Really Works* Smil shows, for example, why the planet isn't 'suffocating' (even burning all the planet's fossil fuels would reduce oxygen levels by just 0.25 per cent) and that globalization isn't 'inevitable' and nor should it be (the stupidity of allowing 70 per cent of the world's rubber gloves to be made in just one factory became glaringly obvious in 2020).

Ultimately, Smil answers the most profound question of our age: are we irrevocably doomed or is a brighter utopia ahead? Compelling, data-rich and revisionist, this wonderfully broad, interdisciplinary masterpiece finds faults with both extremes. Looking at the world through this quantitative lens reveals hidden truths that change the way we see our past, present and uncertain future.

He just wanted a decent book to read ...

Not too much to ask, is it? It was in 1935 when Allen Lane, Managing Director of Bodley Head Publishers, stood on a platform at Exeter railway station looking for something good to read on his journey back to London. His choice was limited to popular magazines and poor-quality paperbacks – the same choice faced every day by the vast majority of readers, few of whom could afford hardbacks. Lane's disappointment and subsequent anger at the range of books generally available led him to found a company – and change the world.

'We believed in the existence in this country of a vast reading public for intelligent books at a low price, and staked everything on it'
Sir Allen Lane, 1902–1970, founder of Penguin Books

The quality paperback had arrived – and not just in bookshops. Lane was adamant that his Penguins should appear in chain stores and tobacconists, and should cost no more than a packet of cigarettes.

Reading habits (and cigarette prices) have changed since 1935, but Penguin still believes in publishing the best books for everybody to enjoy. We still believe that good design costs no more than bad design, and we still believe that quality books published passionately and responsibly make the world a better place.

So wherever you see the little bird – whether it's on a piece of prize-winning literary fiction or a celebrity autobiography, political tour de force or historical masterpiece, a serial-killer thriller, reference book, world classic or a piece of pure escapism – you can bet that it represents the very best that the genre has to offer.

Whatever you like to read – trust Penguin.